Mechanics and Uncertainty

Mechanics and Uncertainty

Maurice Lemaire

Series Editor
Pierre Devalan

iSTE

WILEY

First published 2014 in Great Britain and the United States by ISTE Ltd and John Wiley & Sons, Inc.

ISTE Ltd
27-37 St George's Road
London SW19 4EU
UK

www.iste.co.uk

John Wiley & Sons, Inc.
111 River Street
Hoboken, NJ 07030
USA

www.wiley.com

Library of Congress Control Number: 2014931646

British Library Cataloguing-in-Publication Data
A CIP record for this book is available from the British Library
ISBN 978-1-84821-629-7

MIX
Paper from responsible sources
FSC® C013604
www.fsc.org

Printed and bound in Great Britain by CPI Group (UK) Ltd., Croydon, Surrey CR0 4YY

Table of Contents

Foreword

Industry requires innovation. Innovations are based on technologies, but while we often talk about product technologies (airplane, train, automobile, rocket, satellite, etc.), we rarely discuss the specific technologies involved in the design, testing and industrialization of these products, i.e. the skills involved in product development. This domain has undergone considerable progress in the past decade, with advances in modeling and simulation tools, highly instrumented tests and the use of supercomputers acting as motors for progress.

However, the complexity and multiplicity of the technical domains involved mean that the challenges of high-performance development, producing good results from the outset, are great indeed. Solutions must be found through the use of multiphysics, bringing together domains such as aerodynamics, solid mechanics, thermics and even acoustics. A statistical and probabilistic approach increases the reliability of results, and design methodologies are undergoing constant improvement, as demonstrated by current optimization applications.

The research aspect of design has undergone considerable progress in recent years. However, knowledge of the industrial skills and applications has not developed at the same speed. Significant developments have been made in process modeling for certain procedures, such as forging, casting and deep drawing, but significant work is still required in other domains, including machining, welding and surface treatments. Even more work will be required before suitable models are available for the whole routing for individual articles.

As vice president of the *Association Française de Mécanique* (AFM), French Mechanics Association, I launched a discussion of this situation with colleagues within the association in 2009. My basic premise was this: that while research aspects were undergoing rapid development, production planning was being left behind. We needed to find a way for the groups responsible for these aspects to work together to develop appropriate solutions. My colleagues at the AFM introduced me to a specialist in industrialization research, Régis Bigot, Professor at the Ecole Nationale Supérieure d'Arts and Métiers, Metz, who I met in 2009. We discussed the subject and its scientific and industrial implications at some length, identifying a number of researchers working in this domain from across France; we invited these experts to participate in a meeting at the *Maison de la Mécanique* mechanics center. To my surprise, our invitation was accepted by researchers from seven different laboratories (Metz, Bordeaux, Clermont-Ferrand, Mulhouse, Compiègne, Troyes and Valenciennes), and a number of intense and detailed discussions took place, finally leading to the creation of a focus group which continued to operate over the course of the following two years.

Through the initiative of Maurice Lemaire, a team, whose members are listed in the Preface, set to work on understanding uncertainty, describing the main modeling tools available and exploring the decision-making process as related to this context. The results of their observation are presented in this book, with the aim of guiding those who wish to contribute to develop knowledge and skills in the field of industrialization.

For me, three main points stood out from our meetings:

– We spent a full half-day session discussing the meaning and precise definitions of terms (notably "uncertain" and "uncertainty"); terminology definition is often an important step.

– An increasing awareness of the place of "uncertainty" in the technical domain: in terms of product development, mastery of uncertainty is lost wherever our knowledge of any aspect is incomplete.

– Finally, we were fortunate to be able to present our work to the wider mechanics community at the *Journée de l'AFM* conference in April 2010, a source of great satisfaction for a new cross-disciplinary group.

This book is the fruit of our reflection, and we hope for innovations that will provide new solutions in the complex domains of industrialization.

Jean-Marc THÉRET
President, AFM – 2010-2013
February 2014

Preface

The purpose of this preface is to provide readers with background information on the source of this book and to thank my colleagues involved in bringing this project to fruition.

This book is the result of an initiative launched by the *Association Française de Mécanique* (AFM), the French Mechanics Association. The AFM gathers those involved in the field of mechanics, based on three areas: science, technology and industry.

Over the centuries, mechanical science has developed through the construction of models, which, until the 19th Century, were thought to have considerable predictive validity, as seen in their application to celestial mechanics, where such models provide perfect results due to the infinitely small duration of observations in relation to cosmic time-scales. Later, the concept of perfect deterministic models began to break down, forcing researchers to take a scientific approach to uncertainty.

Mechanical technologies are the result of man's creative abilities, and are, to a certain extent, subject to chance, as their robustness and reliability can never be fully guaranteed. The performance of these technologies is therefore subject to uncertainty, and developers seek the best possible balance between the desired outcome and the cost of potential losses.

The mechanics industry implements technologies for design and production purposes. The design cycle, the production process and the life cycle of a product, as perceived by the user, are also subject to an environment for which the pre-existing information is incomplete. Uncertainty is also present in the components and systems of industrial projects. It can, additionally, come from the organization and its levels of resilience.

The first scientists to experience uncertainty in mechanics were undoubtedly those working on natural forces (such as earthquakes and storms), followed by those working on natural materials, such as soil. The awareness of uncertainty in structural mechanics and in design and production came later, and researchers needed to find ways to respond to the need for robust and reliable designs which go beyond the simple, but essential, manufacturing quality control.

Through the initiative of the AFM, I was invited to participate in a working group on "designing for robust production", led by Jean-Marc Théret and Régis Bigot, which brought together researchers from across the field of mechanics. The group identified vocabulary as a primary reason for uncertainty, both in methodology and in the description of objectives. For this reason, the decision was taken to create a document to precisely specify these concepts, introducing readers to a variety of methods, their strengths and their limitations. In the course of our discussions, participants were encouraged to explore and explain their thoughts in as much depth as possible, providing contributions that were then discussed and integrated into a homogeneous report.

I wish to thank these participants (in order of their appearance in the text):

– Jean-Yves Dantan for his remarks on sections 1.1.1 and 2.3.2.2.3, and his contribution to section 1.3, entitled *Designing for robust production*, through discussions with Régis Bigot and Alain Etienne.

Jean-Yves Dantan and Régis Bigot are Professors and Alain Etienne is an Assistant Professor at the Ecole Nationale Supérieure d'Arts et Métiers in Metz.

– Thierry Yalamas for his contributions to section 1.3.5.

Thierry Yalamas is an Associate Director of Phimeca Engineering S.A. in Paris.

– Nicolas Gayton for his contributions to discussions of section 1.2, entitled *Robustness and reliability*, section 1.3.3.1 concerning capabilities and contributions to the section on *Kriging* methods (i.e. section 2.3.3.2).

Nicolas Gayton is an Assistant Professor and research supervisor at the IFMA in Clermont-Ferrand.

– Sébastien Castric for his contribution to discussions of section 2.1.2.2 concerning *quantitative approaches to random uncertainty* and for his contribution to section 2.2, entitled *Uncertainty in behavior models.*

Dr. Sébastien Castric is a visiting researcher and Assistant Professor at the UTC in Compiègne. He has recently taken on training responsibilities for Airbus.

– Sébastien Berger for his contributions on *polynomial chaos*, sections 2.1.4 and 2.3.2.4.

Sébastien Berger is a Professor at the INSA's Val de Loire center in Blois.

– Felipe Aguirre for rereading section 2.1.2.2 on *quantitative approaches to random uncertainty* and sections 2.3.2.2.1 and 2.3.2.3 on *Possibilistic models*.

Dr. Felipe Aguirre is a research and development engineer at Phimeca Engineering S.A. in Paris.

– Cécile Mattrand for rereading section 2.2.2.5.1 *on Markov processes*.

Cécile Mattrand is an Assistant Professor at the IFMA, Clermont-Ferrand.

– Jean-Marc Bourinet for his contribution on *support vector machines (SVM)*, section 2.3.3.1.

Jean-Marc Bourinet is an Assistant Professor at the IFMA, Clermont-Ferrand.

– Vincent Dubourg for his contributions on *Kriging methods*, section 2.3.3.2.

Dr. Vincent Dubourg is a research and development engineer at Phimeca Engineering S.A. in Clermont-Ferrand.

– Jean-Luc Dulong for his contributions on the question of *model reduction*, section 2.3.4.

Jean-Luc Dulong is an Assistant Professor at the UTC, Compiègne.

– Yann Ledoux and Patrick Sébastian for their contributions to section 3.1, entitled *Decision support in design*.

Yann Ledoux is an Assistant Professor at the University of Bordeaux-1.

Patrick Sébastian is an Assistant Professor and research supervisor at the University of Bordeaux-1.

– Jean-Marc Théret for his contributions to the conclusion, section 3.2.1.2.6, concerning *the TRL scale*.

Jean-Marc Théret is the vice-president of Messier-Bugatti-Dowty, Paris.

The beginning of each contribution is indicated by a footnote and the end is marked by the symbol □ in the text.

I also wish to thank my colleague and friend André Lannoy of the Institut de Maîtrise des Risques (IMdR) for his valuable contributions in rereading and commenting on the preparatory document.

This list is not exhaustive, but it is sufficiently representative to show the conceptual and methodological aspects of uncertainty in mechanics. We hope that this book will help those involved in mastering uncertainty in mechanics to position their research and other activities within the broader, multidisciplinary context of the domain.

Maurice LEMAIRE
February 2014

Introduction

The title of this book "Mechanics and Uncertainty" joins these two terms, which are rarely encountered together. *Mechanics* is "a science, a technology and an industry" focusing on the study of movement, deformation and the states of equilibrium of systems. *Uncertainty* is a philosophical concept, often associated with questions concerning the nature of mankind and our destiny, but also with human creations. The convergence of science and uncertainty constitutes an acceptance of the impossibility of a deterministic predictive model, and our obligation to consider mechanical science in relation to uncertainty.

Here we will consider the history of our topic and the contributions made by a number of great thinkers, before providing a detailed overview of the contents of this book.

I.1. Mechanics and uncertainty throughout history

Confronted with the risks generated by natural disasters and by its own technical innovations, mankind began by turning to the gods. In 1722 BC, Hammurabi promulgated a code establishing sanctions of an eye for an eye, a tooth for a tooth in cases where users fell victim to construction faults. Over time, progress in reasoning, and in the calculation and observation of the natural world resulted in the replacement of the sanction principle by the idea of the predictability of events, modeled by mathematical algorithms, and the definition of their acceptability. In 1609, for example, Kepler published his famous set of laws establishing the elliptical trajectory of the planets around the Sun. In 1638, this was followed by Galileo's discourses on two new sciences, which constituted an introduction to modeling in mechanics. The success of these theories led many thinkers to believe that nature could be reduced to a set of mathematical expressions by the construction of increasingly precise models, and works continued in this regard until a point, at the end of the 19th Century, where science appeared to be more or less "finished".

Gerolamo Cardano (1501–1576), a renaissance mathematician, mechanical thinker and professional gambler, laid the foundations for probability theory in a book named *Liber de ludo aleae*. Later, Blaise Pascal discussed the role of chance in gambling in 1654, introducing the concept of chance as a factor in games: *when we work for tomorrow, and do so on an uncertainty, we act reasonably, for we ought to work for an uncertainty according to the doctrine of chance which has been demonstrated.* He invented the term *geometry of chance*, meaning that chance exists, but that it has a structure; this came to be known as the probability theory, i.e. proof theory, although it does not, technically, prove anything. As early as 1776, Buffon made the connection between statistics and reliability in considering the performance of wooden beams.

For mechanical engineers working in product design, the first of the two themes discussed above led to what may be called the *deterministic fallacy*. The second theme was largely ignored for many years, except in the creation of safety factors. Probability theory also introduced interpretation errors, or the *ludic fallacy*, as it only operates perfectly for games with perfectly defined rules.

At the beginning of the 20th Century, Henri Poincaré (*Sciences et Méthodes*, 1903) broke the deterministic certainties of models of the solar system, which had previously appeared to be perfect due to the infinitely short period of observation in relation to the real durations involved. The deterministic fallacy was demonstrated by the sensitivity of models to initial conditions: *a very small cause, which escapes us, produces a considerable effect that we cannot avoid seeing, and then we say that this effect is due to chance.* This reflection led to the idea of deterministic chaos, but Poincaré did not move outside of the framework established by Laplace: *nothing would be uncertain for an intelligence possessing perfect knowledge of all of the forces of nature.* Von Neumann and Wiener were the first to move beyond this limitation, with the idea of stochastic chaos. In engineering, the acceptance of chance, even framed chance, was subject to major opposition, as Le Chatelier declared in 1924 that *chance offers an escape route to the incompetent, who shy away from taking a scientific approach.*

However, engineers considering risk levels introduced statistical notions, following Mayer (1926) for construction, as part of the expertise process for design variables. A. Grenville Pugsley introduced a requirement concerning accident rates for aircraft in approximately 1930, and others, such as R. Lévi (1949) and A.-R. Rjanitzyne (1949), made use of the laws of probability. A.-M. Freudenthal (1949) *placed the concept of safety of structures in the realm of physical reality, where there are no absolutes and where knowledge is not perfect.* Noting the random distribution of initial defects, Weibull (1951) considered ruptures within a probabilistic framework, paving the way for the notion of mechanical reliability.

This context, coupled with the computing power now available to us, formed the basis of research activities in the second half of the 20th Century, leading to significant developments in probabilistic mechanics and its application to reliability and risk analysis. It also formed the basis for reflection by French engineers such as J.-C. Ligeron and C. Marcovici (1974) or A. Villemeur (1988), who considered ways of predicting the operational safety of electronic and mechanical components.

The work of A.-N. Kolmogorov (1903–1987) sparked the development of probability theory as a rigorous mathematical tool, which was then taken up by economists; however, the ludic fallacy is clearly visible in such cases, as the rules of the "game" are even less well known in economics than in mechanics.

The Mechanical Sciences and the Industrial Future of France (*Les sciences mécaniques et l'avenir industriel de la France*), published by the French *Académie des Sciences* in 1980 [INS 80], called for the creation of scientific teams bringing together experts in mechanics and probability. Thirty years later, the wishes of the *Académie des Sciences* have still not been completely fulfilled, and work is still required in this area; however, it is important to note that mechanics and probability alone are not sufficient. The issues involved in design, operation, maintenance and durability are subject to levels of uncertainty where probability alone cannot provide satisfactory responses in terms of risk management, whether in technical, economical or societal terms.

Mechanics and probability are clearly connected in the standard physical model, but this does not constitute an observational scale as required by the kind of mechanics discussed in this book. The question of the existence of chance is a matter for philosophical debate. In this case, we would do well to follow the ideas set out by Euler rather than Lagrange. In the context of the 21st Century, we are faced with a series of events that appear to be the result of chance, as it is impossible to reconstruct their trajectories. A scientific approach to uncertainty in mechanics needs to be developed.

I.2. Aims and outline of the book

Researchers are increasingly aware of the need to situate their work in an uncertain context, i.e. within a framework defining an imperfect knowledge frame, or, as Pascal put it, a geometry of chance. Technologies are never perfectly robust or reliable, and industrial products are subject to a variety of random situations both in creation and in implementation. Mathematical, physical, technological and industrial perspectives have led to the creation of a variety of methods and procedures, each contributing to a better understanding of uncertainty. The aim of this book is to summarize this information, enabling researchers to position each partial contribution within the broader scientific approach to mastering uncertainty.

For this reason, we will consider a large number of concepts, some at a relatively superficial level, highlighting their contributions and perspectives for implementation. The bibliography given for each point is not exhaustive; the main references will be cited, with a focus on French contributions, showing the richness of the work carried out at national level.

This book contains few mathematical demonstrations and is intended to provide an accessible starting point for scientists, engineers, industrialists and students interested in including aspects of uncertainty in their work; uncertainty is inherent to our world, and the consideration of this notion provides both strategic and operational benefits.

This book is divided into this Introduction and three chapters including eight sections, exploring uncertainty in mechanics following the order as shown in Figure I.1.

I.2.1. *Chapter 1: Understanding Uncertainty*

Developing an understanding of uncertainty involves consideration of three aspects: accepting uncertainty, defining measures and creating an approach to uncertainty.

– Section 1.1. (Uncertainty and reality):

This section will deal with historical considerations in greater detail and highlight the different ways of approaching uncertainty based on the available information, clearly showing the insufficiency of the probabilistic model.

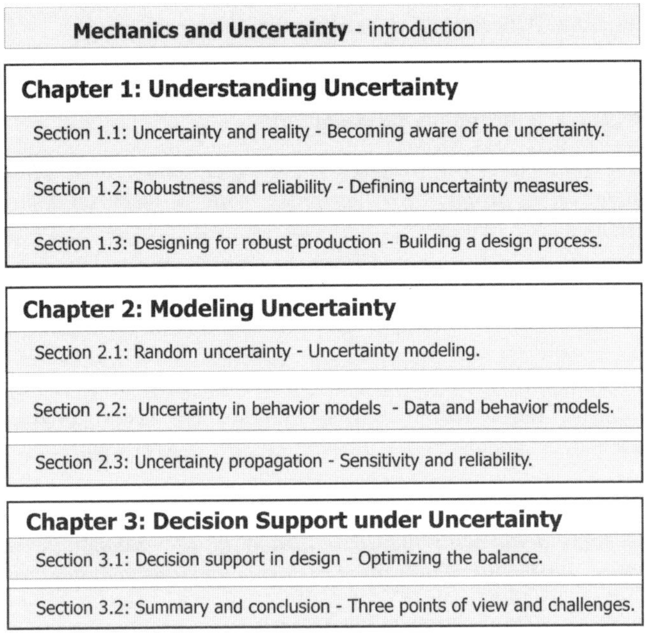

Figure I.1. *Exploring uncertainty: understanding, modeling and decision support*

– Section 1.2. (Robustness and reliability):

In this section, we will discuss the terms "robustness" and "reliability" – two facets of the same requirement for product quality – and the means of measuring these criteria.

– Section 1.3. (Designing for robust production):

This section analyzes the phases and steps in the design cycle and the approach to mastering robustness and reliability.

I.2.2. *Chapter 2: Modeling Uncertainty*

Uncertainty is modeled by representing data and behavior models and the propagation of data within models.

– Section 2.1. (Random uncertainty):

In this section, we examine the way in which knowledge is used to create representative models of uncertainty, from epistemic to aleatoric.

– Section 2.2. (Uncertainty in behavior models):

This section discusses the way in which data and behaviors are modeled.

– Section 2.3. (Uncertainty propagation):

This section details the propagation of uncertainty, in relation to the advancement of a project, for variables with or without measures of probability.

I.2.3. *Chapter 3: Decision Support under Uncertainty*

Decision support in the context of uncertainty necessitates compromises between design requirements and satisfaction for users, engineers and researchers.

– Section 3.1. (Decision support in design):

This section deals with the optimization of the balance between the contradictory constraints of industrial performance, robustness and reliability.

– Section 3.2. (Summary and conclusion):

In this section we will first provide a brief overview from the point of view of the users, the engineers and the researchers. We will then conclude with a discussion of different perspectives for the development of a science of uncertainty in engineering.

Chapter 1

Understanding Uncertainty

1.1. Uncertainty and reality

Uncertainty is inherent to real life, whether it be natural or the result of human activities. Mankind has long been aware of the need to master this uncertainty; however, this awareness does not always lead to the development of a tried and tested methodology, particularly in the domain of mechanics. All too often, the existence of models that are entirely satisfactory in a well-established theoretical framework hides an inability to link these data or behavior models to existing information. Take, for example, the traditional notion of "safety" factors that, in simple terms, are the result of the expert analysis of a situation including unknown factors, without explicit description of the contents of this expert analysis. In this section, we will consider the necessity of developing awareness of uncertainty, highlighting the limitations of current knowledge and proposing a classification of uncertainty, which will be developed in the following sections.

1.1.1. *Awareness of uncertainty*

From the moment mankind became aware of the capacity for learning, we have been interested in the nature of observed events and the prediction of future events. Moving beyond animal reflexes, conditioned by the correspondence between the internal clock and the astronomical clock, which creates awareness of daily and seasonal cycles, human beings were able to contemplate the various events that interfered with these cycles: events that were considered to be uncertain – i.e. the results of chance or an accident – vague, little known or unknown. Our capacities for observation and reflection then came into play with attempts to understand and predict the rules subjacent to all non-ordinary events. Any deviation from predictable behaviors was attributed to chance. This led to a new question: does chance have a cause? The first answers came from religious sources, with the idea that the gods might find it amusing to interfere with human life, or to send down trials as punishment. The way to manage uncertainty, therefore, was through prayer and sacrifices, intended to appease the wrath of the gods.[1] The earliest philosophers, beginning with Socrates and Plato, considered the nature of knowledge, distinguishing between "visible" and intelligible knowledge. "Visible" knowledge, i.e. the knowledge accessible to the senses, is based on conjecture and conviction, whereas intelligible knowledge is based on science and believed to be genuine. Over the course of time, rational explanations emerged as a result of the human capacity to understand and explain phenomena. In the 18th Century, however, Emmanuel Kant expressed a concern that, however it might be expressed, science could never fully reflect real life. As our knowledge of the mysteries of nature progresses, we become increasingly aware of the limitations of our knowledge as it stands □. Kant's warning went unheeded by 19th Century scientists. Pierre-Simon de Laplace believed in causal determinism, as expressed in the introduction to his Philosophical Essay on Probabilities:

> *An intellect which at a certain moment would know all forces that set nature in motion, and all positions of all items of which nature is composed, if this intellect were also vast enough to submit these data to analysis, it would embrace in a single formula the movements of the greatest bodies of the universe and those of the tiniest atom; for such an intellect nothing would be uncertain and the future just like the past would be present before its eyes.*

Probability is only used to counteract gaps in our knowledge, and epistemic progress should lead to precision in predictions. At the end of the century, scientists were able to announce the completion of the "scientific conquest", with the exception of certain small details; early in the next century, however, these "small" details proved to be rather more significant than was first thought. At the end of the 19th Century, physics was based on

1 From a contribution by J.-Y. DANTAN [DAN 09].

two pillars: Newtonian mechanics, and Maxwell's ideas on electromagnetism [KLE 08]. Each theory appeared to be correct, but their principles were incompatible. The downfall of the 19th Century scientists was spectacular, as Henri Poincaré remarked in the early 20th Century with the introduction of the notion of chaotic behaviors: *A very small cause which escapes our notice determines a considerable effect that we cannot fail to see, and then we say that the effect is due to chance*. This reflection was the starting point for the idea of deterministic chaos, but Poincaré did not overstep the boundaries of the framework established by Laplace. The *deterministic fallacy* was highlighted by a sensitivity to initial conditions that can never be known with a sufficient level of precision. This error was made increasingly apparent in the work of John Von Neumann and Norbert Wiener on trajectories subject to noise, which led to the development of stochastic chaos as a tool for predicting reality. Deterministic calculation was finally condemned and limitations to reasoning were established by Kurt Gödel's incompleteness theorem. Nowadays, we know that our information is, and will always be, incomplete, proving a point made by Emmanuel Kant: *Someone's intelligence can be measured by the quantity of uncertainties that he can bear*. The acceptance of uncertainty in technologies was difficult, as expressed by Henry Le Chatelier, founder of the modern French chemical industry, in the early 20th Century [BAY 95]: *the hypothesis of chance offers an escape route to the incompetent, who shy away from taking a scientific approach*. However, uncertainty is real; it can be mapped, and this map can be explored using a scientific approach.

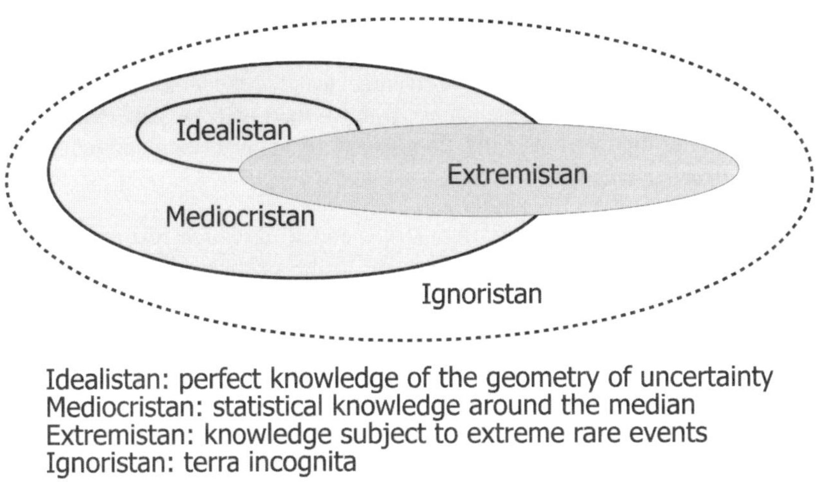

Figure 1.1. *Idealistan, Extremistan, Mediocristan and Ignoristan territories*

1.1.2. *Territories of uncertainty*

As a concept, the notion of the *geometry of chance* owes its existence to Blaise Pascal. "Chance" may be said to exist when an event has a random outcome. The geometry of chance has, regrettably, been absorbed into probability theory, i.e. proof theory; a bijection should not be made between chance and probability. Chance has a structure, and under certain conditions, this structure may be represented using probability theory. In such cases, it is important to identify the structure. Even before Pascal, all who thought about chance in any depth linked the notion to its implication in games. Games are a human invention, with rules and a set of events for which a probability space may be easily identified. These considerations mean that a probabilistic modeling is relevant in such cases. The generalization of these methods to abstract sets led to the perfect mathematical establishment of probability theory by Kolmogorov. However, while all is perfect in theory, it is impossible, with the exception of academic examples or game theory, to construct a geometry of chance. In mechanics, we are faced with a need to master uncertainty using information that must always be insufficient. Uncertainty is an intrinsic part of real life, and the structure of uncertainty is, in itself, uncertain.

Taleb [TAL 07] referred to the probabilistic model of economic chance as the *ludic fallacy*; as in the case of mechanics, this model of chance must always be based on incomplete information. Taleb established a distinction between two territories of uncertainty. The first, "Mediocristan"[2], refers not to a situation of mediocrity but to the median. In this territory, all events are located around the median, often close to the mean. In this case, new observations do not lead to significant modifications to acquired knowledge. Examples of this type include the average weight of the inhabitants of a country or manufacturing dimensions; at most, there will be slow evolutions over time. In this territory, the dominance of statistics and probabilistic prediction is unchallenged.

The second territory is "Extremistan", and in this area, extremely rare events can create significant modifications to the parameters of uncertainty. This is the case, for example, for seismic levels in France, established based on the background noise resulting from observations, which may be subject to sudden changes in the future. We wish to know whether the noise perceived

---

2 Mediocristan refers to that which is located around the mean or the median. The term "Medianistan" may be considered preferable, as it eliminates the negative connotations of "mediocre"; however, we will retain the original vocabulary selected by Taleb.

over a period of several years can be extrapolated to a sufficient level. These considerations are dealt with in the domain of extreme statistics and in connection with risk: the consequences of passing a certain level are, however, seen to be acceptable by society or by individuals. In the last case, the mastery of uncertainty involves the decision to accept or reject the rare event, and not through a probabilistic hypothesis, which would need to be compared to that used for other events of the same type, which are not taken into account – such as the probability of a meteorite or satellite falling to Earth at a given location. The probability that a satellite, at the end of its lifecycle, with a trajectory that is completely unpredictable a few hours before falling, will fall on a given surface of 1 km^2 is 1.96×10^{-9} (the inverse of the area of the globe).

Taleb's mapping may be supplemented by the addition of "Idealistan", a territory where our knowledge of the structure of chance is perfect, inhabited by game theory and by designers who believe that reality is presented on a computer screen, and "Ignoristan", inhabited by all unimagined events. In this case, we need to use ignorance exploration techniques, such as failure mode, effects and criticality analysis (FMECA), functional analysis, quality assurance and any number of imaginable tools, as discussed by Ligeron [LIG 06]. Figure 1.1 illustrates the territories of uncertainty where a form of classification may serve as a guide.

Thunnissen [THU 03] explored the perception of the notion of uncertainty in a large number of disciplines as seen in the published literature. He highlighted major differences in understanding according to domains, producing a classification that will be discussed below.

The first cause of uncertainty is *ambiguity*, which is the result of a lack of precision in language, in a project, in design and in feedback, where descriptions are often too vague. Words are random variables for which the common meaning differs between different communities and is subject to deviation. Uncertain ambiguity should not be confused with the reasoned choice not to provide precise values during the design phase, so as to allow a certain degree of freedom. Within the aeronautics industry, a simplified language (*simplified English*) (http://en.wikipedia.org/wiki/simplified_english) was created in order to guarantee perfect understanding in all areas, particularly for maintenance purposes. From a formal perspective, the theory of fuzzy logic allows us to attribute a degree of trust to an affirmation when dealing with ambiguities.

The second cause of uncertainty results from the *random* or *aleatoric character* of certain information. The word "random" arises from the French term "root", which means "to run wildly", generally used for horses; it is the equivalent of the French term "aléatoire", from a Latin word referring to the game of dice. The word "stochastic" arises from a Greek term, which means "soothsayer".

Aleatoric uncertainty (or stochastic uncertainty, used as an equivalent term) concerns variations inherent to a physical or mechanical system and its environment. Is this uncertainty real, or simply the result of our ignorance? It is intrinsic to the quantum model, as stated in [AGU 13], but in mechanical engineering, this uncertainty is purely the result of our inability to access and use all data.

This uncertainty is therefore impossible to control, and constitutes a background noise, which may vary within a domain of more or less fixed boundaries. Expert opinions or physical considerations help us to set these limits, and probability theory provides a structure for the distribution of variations, a structure that we then need to identify. Random uncertainty is irreducible and objective. The dimensions of pieces during fabrication and the properties of materials are examples of this kind of uncertainty. Probability theory produces a perfect model of random knowledge, but the structure of this random aspect remains approximative, except in the case of game theory. "Aleatoric" is not, therefore, necessarily equivalent to "probabilistic".

Epistemic uncertainty concerns any lack of information, which may be remedied by measurements, appropriate actions or decisions. It is therefore reducible and subjective, and may be classified in relation to two main sources:

– The *stochastic modeling of design variables*: each variable possesses a description that depends on the level of knowledge. This depends on the advancement of the project: "the material is steel", and on the depth of physical knowledge: "its characteristic elastic limit is 240 MPa". According to the advancement of the project and the scale of observation, it may be represented using different entities (section 2.1).

– The *physical modeling of behaviors*: the description of a phenomenon using a mathematical model benefits from a rigorous and coherent approach, but involves the use of hypotheses that may be more or less well validated. The solution to the model, which is generally numerical, only gives an approximation of the solution and may even include errors in the programming of the algorithms. While errors may be avoided by careful quality assurance, the representation of gaps (and not errors!) between the model and the physical reality remains a subject for research. Recalibration methods are most effective on a case-by-case basis, but they do not allow us to map gaps across the domain of validity. The lack of phenomenological knowledge is also a source of epistemic uncertainty, and plays a particularly important role in cases of innovation in a mechanical system or in a new environment. In these cases, the designer only has access to subjective estimations (section 2.2), and we should remember that the behavior may fall within the territory of Extremistan.

The solution to epistemic uncertainty, i.e. the production of a certain representation, is clearly impossible in cases where an uncertainty is represented by a continuous variable in a domain, even if the domain is bounded. In this case, the inclusion of successive information allows us to construct a model with parameters that are themselves tainted with random uncertainty, and the introduction of new information causes the process to repeat itself. The representation of epistemic uncertainty thus appears to be fractal and always results in random uncertainty. Epistemic uncertainty would, however, be solvable if (following Laplace) a higher form of intelligence proved able to acquire the information needed for exhaustive knowledge of the space of possibilities; that being said, Poincaré reminds us of the issue of sensitivity to initial conditions. One example is provided by clearance mechanisms, where a random drawing selects one of a finite number of identifiable combinations [DAN 12].

Other sources of uncertainty exist, which will not be examined in depth in this book. *Volitional uncertainty*[3] results from human behaviors over which the designer has little or no control when specifying possible choices. It concerns design choices where the exhaustiveness of solutions has not been reached, the requirements that individuals may introduce independently, the way in which the will of the user may be expressed in given conditions and crude errors. Finally, *interactive uncertainty* concerns the simultaneous occurrence of certain events or disciplines, or potential "forking" from one behavior to another. If we accept that a single, clearly identified cause almost never leads to a highly undesirable event, as its simple identification leads to protection measures being taken, extreme events are generally the result of multiple causes, none of which would have extreme effects if taken separately.

The distinction between aleatoric and epistemic uncertainties tends to disappear when uncertainty is represented by a model. If an uncertainty is

3 Volition is an act in which something is determined by the will. In other terms, it designates the finality of a process where a human being makes use of will. This is the event by which an individual is "able to act" in relation to an internal or external result. In common parlance, volition may be assimilated to the formulation of a choice (http://en.wiktionary.org/wiki/volition).

aleatoric and represented by a chosen probability density function (PDF), then no additional information should be able to change this choice, even though it was chosen based on incomplete information. The choice of a PDF is therefore a matter of epistemic uncertainty, in which we should include physical knowledge, for example whether or not the density is bounded. We also need to include the results of the composition of uncertainties: Weibull's law for materials, for example. Epistemic uncertainties may be represented by probability densities when sufficient information is available to define the first statistical moments. In this case, the parameters of the densities are known and a probabilistic model may be chosen. As highlighted in [KIU 09], the distinction between epistemic and aleatoric uncertainties is a result of choices made during the modeling process.

Finally, for any type of uncertainty, modeling converges toward a PDF; this attitude may be compared to placing the mastery of uncertainty under the "light"[4] of probability theory. Failing that, we need to be able to master heterogeneous information combining cognitive, heuristic and algorithmic knowledge.

1.1.3. *Conclusion*

The awareness of uncertainty leads to recognition of the *deterministic fallacy*. Uncertainty is inherent in life and in mechanics, and it must be mastered. To do this, we need to clearly distinguish between situations: those where we have a large amount of information – which belong to Mediocristan and for which we have a probability map – and those that result from rare events, or even extrapolations – which belong to Extremistan – with a corresponding risk map. A situation is uncertain if the outcome may take different forms, which may or may not be identified within the probability structure that defines the associated random variables. While this theory defines a perfect mathematical framework based on a stochastic data model, exhaustive knowledge is never possible as mechanics is not subject to the strict rules of game theory, nor do we possess samples of the size used in opinion polls, for example. Avoiding the *ludic fallacy* is significant. The outcome of an uncertain test is therefore represented based on the available

4 The Lost Key: Night time, in a street, near a street lamp. A man is standing, looking at the ground, and appears to be looking for something. Another man walks by and asks: "What are you looking for?" – "I'm looking for my key". – "You've lost your key?" – "Yes". – "You lost it here?" – "No". – "But if you lost it somewhere else, why are you looking here?" – "Because there's light here" [CAR 98, p.354].

knowledge using different models, which we will examine in 2.1, whose limit may be a random variable or stochastic process.

The following sections will be devoted to the notions of robustness and reliability, and to the consideration of uncertainty within the design process. We will then go on to explore the territories of uncertainty in more detail (sections 2.1 and 2.2).

1.2. Robustness and reliability

A clear understanding of uncertainty is essential for mechanical engineers who wish to develop robust and reliable products.[5] In this section, we will analyze the relationship between uncertainty, robustness and reliability, attempting to clarify issues of vocabulary, which constitute a primary cause of uncertainty. We will then consider the question of optimization.

1.2.1. *Robustness*

1.2.1.1. *Definition and measure of robustness*

In the popular meaning, robustness is the quality of something that is robust and the word "robust" is associated with strength: that which has the necessary strength to resist events is robust. Antonyms include words such as weak and feeble: to be in a robust state of health is to have strong resistance to uncertain potential attacks by disease. A design is robust if it has the capacity (the strength) to resist uncertain and/or unexpected events that the object may encounter in the course of its lifecycle. The word "robustness" takes on different meanings for different disciplines (http://en.wikipedia.org/wiki/robustness): robust algorithms in numerical analysis converge toward the right solution in all circumstances; robust estimators in statistics converge to the desired parameter as the size of the sample tends toward infinity; and in mechanics, robustness refers to the solidity of a material or product (although this notion is more closely connected with the idea of reliability). Robustness also refers to *performance stability in engineering*. We will retain this latter perspective, as it characterizes the quality of adaptation of a product or performance to chance factors, highlighting its stability, but not necessarily its optimality. We will therefore adopt the following definition:

---

5 Subject discussed with N. GAYTON.

In an engineering context, a robust design is one that determines the nominal parameters of a product or a system so that its performance is sufficiently insensitive to any uncertain event, which it may encounter in the course of its lifecycle.

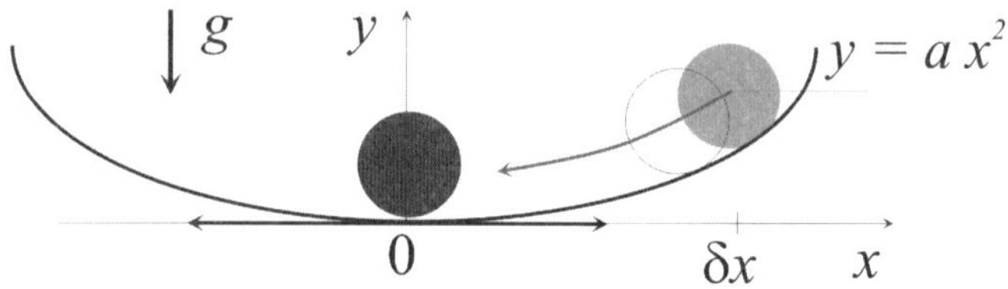

Figure 1.2. *Illustration of the principles of robustness: ball bearing in a cup*

This definition is simple to understand in the case of dynamic systems, where any perturbation will create an opposing force drawing the state back to the point of equilibrium (e.g. a ball bearing in a cup, an automobile suspension system, the elastic stability of shells or control systems). It is more difficult to understand in a static context. Two different modes of robustness coexist. The first mode is linked to the behavior model (*robustness of behavior*) and the second mode is linked to the nominal parameters of the system (*robustness of parameters*). Let us take the example of a ball bearing in a parabolic cup. The position parameter is x, which implies a performance $y = a\,x^2$. Clearly, any solution $x \neq 0$ will be unstable in the gravitational field g as the mechanical behavior of the system leads to a return to the initial position following any variation δx, unless δx exceeds a threshold where the ball leaves the cup. This shows that robustness may be limited by the variation of parameters.

The formal expression of this property is based on the notion of virtual variation. Let us consider a set of design parameters $X_1, X_2, ..., X_n$, a *response* defined by a *variable of interest* $Y = \mathbf{M}(X_1, X_2, ..., X_n)$ and $(\delta X_1, \delta X_2, ..., \delta X_n)$, an admissible virtual variation of the design parameters. Thus:

$$Y(X + \delta X) = Y(X) + \sum_i \left.\frac{\partial \mathbf{M}}{\partial X_i}\right|_X \delta X_i + \frac{1}{2}\sum_{i=1}^{n}\sum_{j=1}^{n} \left.\frac{\partial^2 \mathbf{M}}{\partial X_i\,\partial X_j}\right|_X \delta X_i\,\delta X_j + ...$$

The first term of the expansion expresses the level of the response. To know whether or not this is sufficient, we need to consider the second mode of robustness, and this question is examined in the following section in connection with reliability. The second term (the gradient) represents the *sensitivity* s_i of the variable of interest Y as a function of the parameters X_i:

$$s_i = \left.\frac{\partial \mathbf{M}}{\partial X_i}\right|_X$$

The components s_i of the sensitivity vector, which are dependent on the dimension of the variables, are therefore not directly comparable and it is preferable to introduce the notion of *elasticity* defined by:

$$e_i = \left.\frac{\partial \mathbf{M}}{\partial X_i} \frac{X_i}{\mathbf{M}(X_1, X_2, ..., X_n)}\right|_{(X_1, X_2, ..., X_n)^{(r)}}$$

where $(X_1, X_2, ..., X_n)^{(r)}$ is a point of reference: the mean, the median or the design point associated with the maximum likelihood of the failure [LEM 09]. The elasticity measure is therefore a one-dimensional vector, which depends on the calculation point.

Finally, the Hessian matrix $H_{ij} = \frac{\partial^2 \mathbf{M}}{\partial X_i \, \partial X_j}$ shows stability, whether positively or negatively defined, according to the context. As an illustration, let us consider the case of a given need (or internal strength or stress) S that must be satisfied by dimensioning a resource (or resistance) $Y(X) > S$ (Figure 1.3); the opposite situation is also possible. A performance with zero sensitivity is connected to each of the design points P_0, P_1 and P_2, but only P_0 is robust as the resource increases as a result of a variation δP_0. Matrix H is thus defined as positive; it would be negative in the case of a need with a given resource. Behavioral robustness cannot exist without extrema.

In cases where dimensioning implies zero sensitivity, robust behavior is only present if, for any admissible virtual variation in parameters:

$\forall \delta X_i$ admissible

$$\delta dY = \sum_{i=1}^{n} \sum_{j=1}^{n} \frac{\partial^2 \mathbf{M}}{\partial X_i \, \partial X_j} \delta X_i \, dX_j \left| \begin{array}{lll} < 0 & \text{if} & \text{need} \\ > 0 & \text{if} & \text{resource} \end{array} \right.$$

where dX_j is a differential increment. Note that this criterion is local for small perturbations around the calculation point. If the expression is null, we must look for higher order derivatives. Points P_1 and P_2 give stronger resources and their use helps to satisfy reliability criteria.

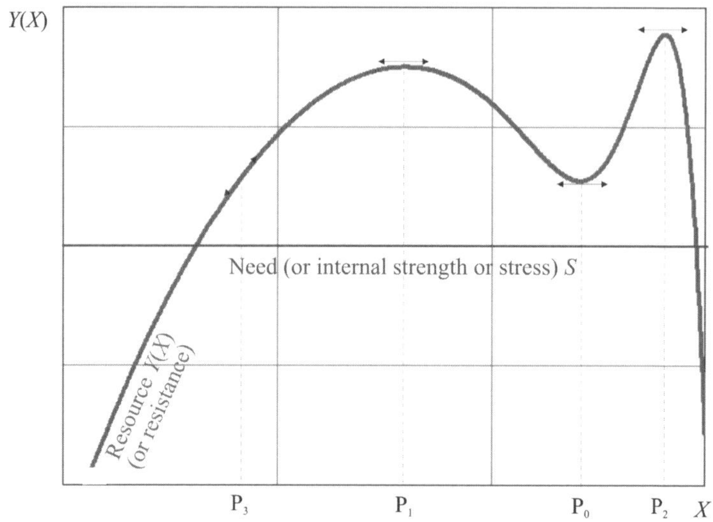

Figure 1.3. *Illustration of robustness – case of a resource and a fixed need*

If performance around a criterion is monotone, then matrix H is defined as neither positive nor negative and is considered to be null in cases of linear behavior. The notion of robust behavior therefore disappears. However, the sensitivity indicates the direction of variation to take toward a better response, as illustrated by point P_3. Thus, the choice of height for a dam may not be robust in behavioral terms, but will simply ensure a certain reliability in relation to the uncertain factor of water levels.

This analysis demonstrates the notion of "*sufficient insensitivity*", introduced in the definition. A design presents *robust behavior* if the sensitivity is such that any perturbation will increase the deviation between resource and need. Robustness therefore depends on the possibility of including non-monotone evolution of parameters in a design. It does not depend on uncertainy, but on *sensitivity analysis*. Where this is not possible, a design will be conditionally robust in connection with identified parameters within a given interval, showing a link with the idea of reliability.

1.2.2. *Reliability*

1.2.2.1. *Definition and measure of reliability*

The term "reliability" expresses ideas of trust and confidence. One definition, used in slightly different forms in a variety of domains, is as follows:

> *Reliability is the capacity of a part of equipment or a product to carry out its function in the required conditions of use over a fixed period of time.*

A quantitative measure is sometimes added to this qualitative definition, referring to probabilities; this will be discussed below.

Reliability is an essential component of reliability, availability, maintainability and safety (RAMS), which includes reliability objectives alongside availability and maintainability criteria in relation to the expected performance, while respecting safety constraints linked to the risks presented by the product to its users or to its environment.

Practical reliability is subject to correct usage procedures, and is associated with quality assurance; theoretical reliability is a means of prediction using mathematical models to estimate the chance of successful performance (demonstration of a resistance function, conformity of production to specifications, allocation of reliability to components in a system, etc.).

The idea of success and failure encourages the use of probability to measure reliability, to the point where certain definitions link the two notions; however, we should remember the importance of *notional probability*, i.e. probability conditioned by the available stochastic model. If these precautions are respected, then probability becomes an essential measure of reliability. In Mediocristan, experimental feedback from relatively large samples may be used to validate the experimental frequency – probability relationship. This is not the case in Extremistan, where feedback is limited to censored occurrences that did not lead to failure. Reliability can thus only be seen as a measure of the state of a system, for which a sensitivity analysis can provide us with information concerning the most influential parameters.

In *reliability analysis*, a variable of interest Y is compared to a threshold $\mathcal{T}$ that may be set at zero. In this case, we are no longer looking at stability, but at a distance from the threshold. The probabilistic measure of reliability is thus the probability of the event:

$$E = \{\text{response } Y = \mathbf{M}(X) > \mathcal{T}\}$$

where Y represents a positive performance. The probability of success becomes:

$$P_{\text{success}} = \text{Prob}\{Y = \mathbf{M}(X) > \mathcal{T}\}$$

and the probability of failure P_f becomes:

$$P_f = 1 - P_{\text{success}} = \text{Prob}\{Y = \mathbf{M}(X) \leq \mathcal{T}\}$$

P_f is often referred to as a reliability measure; however, it is actually a measure of *unreliability*.

In mechanics, a stress S is covered by a resistance $Y(X) = \mathbf{M}(X) > S$, and we use $G(X, S) = G(\mathbf{M}(X),\ S)$ to denote the function such that $G(X, S) > 0$ represents success and $G(X, S) \leq 0$ represents failure. The conditional measure of reliability is thus P_f =Prob$\{G(X, S) \leq 0\}$, given the stochastic model of X and S.

The example in Figure 1.4 illustrates this idea. The stress is presumed to be fixed at a value of S_i for each point, such that we obtain a positive margin $M_i = Y(P_i) - S_i$. Let $f_Y(y)$ be the probability density of the resource at P_i. At P_0, there are no realizations of P_0 that give $G(Y, S_0) < 0$, without looking for an extremely broad distribution. At P_1 and P_3, we have $G(Y, S_{1 \text{ or } 3}) < 0$, and the gray zone illustrates failures, one associated with zero sensitivity the other associated with monotone behavior.

Figure 1.5 shows the respective densities of an uncertain requirement $f_S(x)$ and uncertain resource $f_Y(x)$, and the hatched zone shows the probability of a resistance being inferior to the stress. The notional probability calculation is thus given by:

$$P_f = \cup dP_f = \int_{\mathbb{R}} f_S(x)\, F_Y(x)\, dx$$

where $F_Y(x)$ is the distribution function of Y.

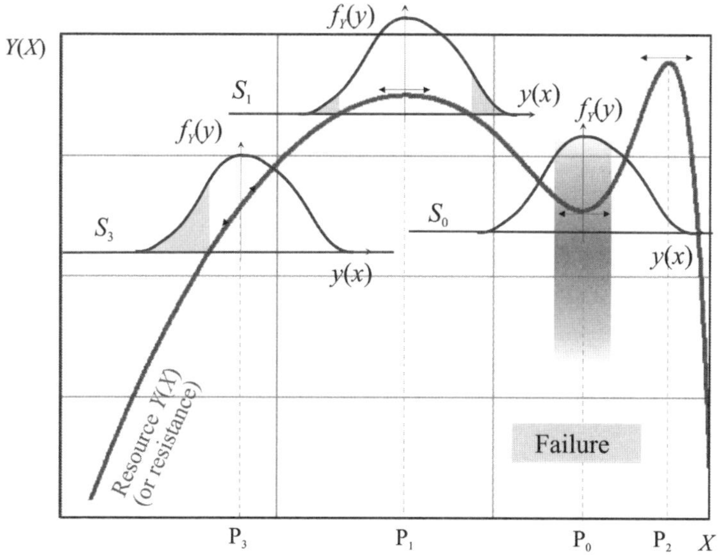

Figure 1.4. *Illustration of reliability in the case of a resource responding to a given requirement*

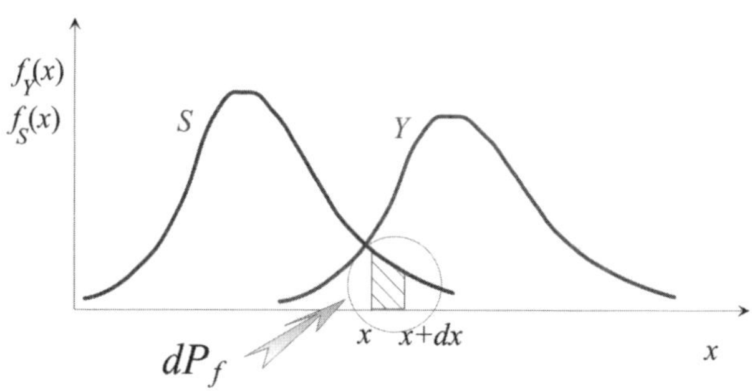

Figure 1.5. *Probability densities of the stress and the resistance*

According to the complexity of the response and the threshold, the calculation of P_f, which is a "simple" integral, will require tailored methods; the direct application of the Monte Carlo simulation [LEM 09] is too expensive in terms of calculation resources.

Although *robustness* is characterized by sensitivity, *reliability* is characterized by the *distance* between the performance function and a fixed threshold. In probability theory, this distance is measured by the probability of failure or by an equivalent reliability index β such that $P_f = \Phi(-\beta)$ where Φ designates the normal distribution with zero mean and variance 1. This index presents the advantage of a logarithmic scale, which is better suited to design sensitivity in relation to risk taking. Other distance measures may be used, the simplest of which uses a safety factor that imposes a deviation between characteristic values of the resistance and the stress □.

1.2.3. *Relationship between robustness and reliability*

1.2.3.1. *Three situations of robustness and reliability*

Let us consider the illustration in Figure 1.6 and examine the solutions as a function of the design points for stresses S_i. We will define two types of parameter variation intervals as follows:

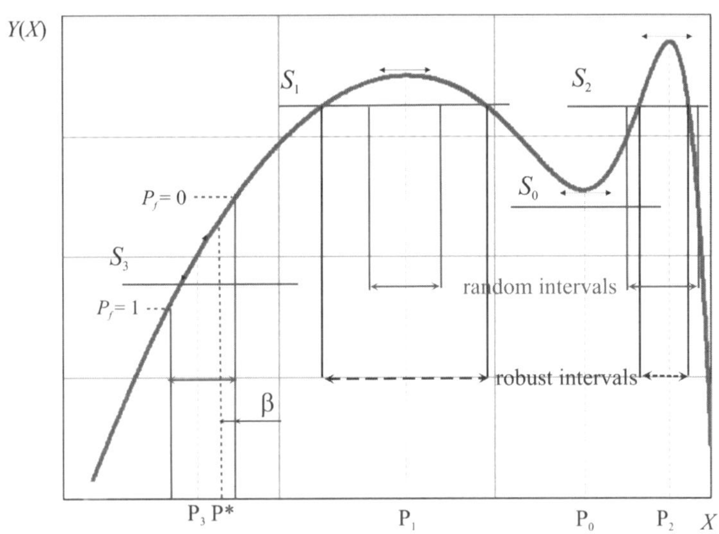

Figure 1.6. *Comparison of robust and reliable situations*

– An interval is said to be robust if the bounds of parameter variations are considered to be certain, without prejudice for the existence of a distribution within this interval.

– An interval is said to be random if it corresponds to the support of the distribution of the parameter.

Let us consider the following three situations represented by a design with points P_i:

– Point P_0 shows the robustness of the ball bearing in the cup. Its reliability is absolute in a large vicinity and the margin is always positive. The design is robust in relation to behavior. Point P_0 is chosen as a dimensioning point and the margin at this point is zero. It represents an ideal robust and reliable situation.

– Point P_1 shows a situation for which large intervals of robustness may be defined. The reliability is total if the random interval is contained within the robust interval. The design is robust in relation to its parameters. The dimensioning point is P_1 and the margin is such that the random and robust intervals are mixed together. The generalization of this approach will be considered in section 2.1.2.3, associating robustness with convexity.

– For point P_2, the robust interval is too limited to contain the random interval; the design is not robust and presents a certain level of reliability. The dimensioning point is P_2 or in the vicinity of P_2 and the margin is defined by the required reliability level.

– Point P_3 is neither robust nor reliable. The dimensioning point P^* is chosen at a distance $\beta = P_3P^*$ in order to give a certain level of reliability. In the figure, if the random interval is bounded, then the dimensioning may be totally reliable.

Robust and reliable design involves two subjacent hypotheses. The first hypothesis requires a behavior model with a positive (respectively, negative) curve, and the second hypothesis requires the existence of a guaranteed interval of robustness. The only remaining uncertainty is thus that of the model. Situations of this kind are extremely rare in practice, and designers must accept a certain level of risk, a level which depends both on the occurrence of the undesirable event and on its consequences.

1.2.3.2. *Reliability and sensitivity*

In an ideal world, we would be able to design in such a way that the resistance, or the stress, would always be a function presenting extrema. In practice, most of the design points we encounter are in the situation of P_3 with an increasing or decreasing slope, measured by its sensitivity s_i, as in the case of robustness.

Distance β is a measure associated with probability. An approximation of this measure is given by the following calculation, which demonstrates the role of sensitivity. Let us carry out the first-order Taylor expansion around the mean m_X of the variables:

$$G(X,S) = G(X,S)|_{m_X} + \underbrace{\left.\frac{\partial G(X,S)}{\partial X}\right|_{m_X}}_{\text{Sensitivity } s_i} (X - m_X) + ...$$

Thus:

$$\text{mean of } G: m_G = G(X,S)|_{m_X}$$

$$\text{variance of } G: \ \text{var}_G = \left[\left.\frac{\partial G(X,S)}{\partial X}\right|_{m_X}\right]^t [\text{cov}_X] \left[\left.\frac{\partial G(X,S)}{\partial X}\right|_{m_X}\right]$$

The first idea of a reliability measure by an index was proposed by Rzhanitzyn [RZH 49], and then further studied by Cornell and Benjamin [BEN 70]. This index is the relationship between the mean and the standard deviation (the inverse of the coefficient of variation):

$$\beta = \frac{m_G}{\sqrt{\text{var}_G}}$$

β is a one-dimensional measure showing the number of standard deviations between the mean (or median) point and the failure point at a given level of reliability, generally a few units, corresponding to the common engineering practice of shifting experimental results by a few standard deviations. It is linked to the probability of failure by the following relationship:

$$P_f \approx \Phi(-\beta) \qquad [1.1]$$

where $\Phi()$ is the normal distribution function. If G is a linear function of X Gaussian variables, then approximation [1.1] is precise. Calculated at a robust optimum point, the variance of G is null as $s_i = 0$ and $P_f = 0$.

This first approach illustrates the connection between robustness and reliability through the use of sensitivity. It was extended to include broader hypotheses [DIT 96, LEM 09], which led to the specification of the sensitivity at the design point (at the point of maximum likelihood of failure, rather than at the mean) and to the specification of the estimator trust interval, i.e. a reliability measure that is robust when faced with uncertainty.

1.2.3.3. *Robustness in systems*

Given the impossibility of designing a robust product based on the behavior of a single component, we must consider the possibility of redundancy, enabling systems to continue to operate, at least at reduced capacity. Redundancy may be active, for example if two elements are responsible for the same function, where one would be sufficient. In the course of normal operations, the two elements receive a relatively small load; in the (highly unlikely) case of failure of one of these elements, the second component would take over the full load. Clearly, these elements need to be independent. Passive redundancy is where a secondary element, in a state of standby, is only activated in case of failure of the primary element.

This idea of redundancy is illustrated by practices in the automotive industry in relation to tire failures:

– Case 1: vehicles are equipped with a spare wheel, which is identical to the wheels used on the vehicle. The wheel is changed and normal operations resume (unless a second tire blows).

– Case 2: vehicles are equipped with a flat spare wheel, which offers reduced levels of performance when compared to the normal situation. The wheel is changed and operations resume at a limited level, i.e. with lower acceptable maximum speeds.

– Case 3: vehicles do not carry spare wheels, but the driver has access to a call center. Operations may therefore be resumed, but only after a certain period of time has elapsed.

In aeronautics, the *fail safe* concept means that a failing element must not have consequences for safety, but not that normal operations must continue. Dimensioning of this type requires the use of redundant effort transmission pathways.

These examples show that the fulfillment of a required function must be subject to a systemic analysis of the consequences of potential failure.

1.2.4. *Optimizing robustness and reliability*

We have now considered robustness and reliability based on performance satisfaction, without looking at notions of performance optimization. Moreover, some only consider a design to be robust if it is optimal, requiring us to measure optimality. Taguchi [TAG 89] worked on this question in the late

1980s, creating a methodology that reduces development time and costs before a product is launched [CLE 00]. This methodology is based on optimizing robustness during design, using data and a measurement obtained through experimentation. Two types of parameters are included in the performance function Y:

– $\{X\}$, vector of the control parameters available to the designer and used for optimization;

– $\{\xi\}$, vector of the noise parameters over which the designer has little influence.

Let $\hat{Y}$ be the target and κ a set of experiments with fixed $\{X\}$. The root mean square deviation measures the dispersion due to the parameters $\{\xi\}$:

$$\text{RMS} = \text{E}\left[\left(Y(\{X\},\{\xi\}_\kappa) - \hat{Y}\right)^2\right] = \sigma_Y^2 + (m_Y - \hat{Y})^2 \approx \frac{1}{\kappa}\sum_{i=1}^{\kappa}\left(Y(\{X\},\{\xi\}_\kappa) - \hat{Y}\right)^2$$

where m_Y and σ_Y^2 are the mean and the variance of Y, respectively. Taguchi proposed the use of the signal/noise ratio as a measure:

$$\text{SN} = -10\log_{10}(\text{RMS})$$

where the optimal robust design is the solution $\{X\}^{opt}$ that maximizes SN with noise $\{\xi\}$.

A key contribution made by this method is the solution of the optimization problem using a carefully chosen range of experiments. While Taguchi's focus on the most robust design, rather than simply robust design, is commendable, the value of the SN measurement is debatable. Moving beyond this experimental method, Beyer and Sendhoff [BEY 07] highlight the need to develop a measure of robustness including representations (deterministic, probabilistic and possibilistic) and tools (worst case, stochastic calculation and genetic algorithms), which we will discuss in greater detail in the following chapter. The authors also highlight the fact that the maximization of performance at the point of design and the minimization of variance across a wide domain of operating conditions often constitute two contradictory aims.

The two keys to robust optimization are the choice of a function measuring robustness, for example the signal/noise relationship or the reliability index, and the means of solving the optimization problem, by experimentation or by algorithms such as those identified by researchers in reliability-based design optimization (RBDO) [ENE 94b, KHA 02, CHO 07]. Algorithms without tests, however, are as limited in their value as tests without models.

1.2.5. *Conclusion*

A product is robust if it is sufficiently insensitive to all uncertain events. It is only totally robust if the dimensioning point is situated in a particularly favorable version of a behavior model, which it is not always possible to design. Otherwise, it can only be totally robust if the parameter definition intervals are absolutely certain. *The essential measure associated with robustness is sensitivity.*

A product is totally reliable if the undesirable event is impossible, either because the product is totally robust or because the random intervals are contained within the intervals of robustness. Otherwise, the product will be reliable to a certain level, generally measured using a notional probability. *The measure of reliability is associated with a distance.*

Table 1.1 gives a summary of possible situations. It highlights the importance of the behavior model for robustness and the uncertainty model for reliability. The notions of robustness and reliability, illustrated here by the case of a resource needed to balance a requirement, must be extended to all functions of design variables, including estimators such as probability itself and the statistical moments of variables.

		Robustness	**Reliabilty**	
Behavior with minimum (zero sensitivity) $s = 0$		Total	Perfect	Insensitivity to uncertainty, except that of the behavior model, margin null
Behavior with maximum (zero sensitivity) $s = 0$	Robust interval containing the random interval	Around the design point	Perfect	Requires certainty of the robust interval, margin defined by the robust interval
	Random interval containing robust interval		Reliability measure	Requires knowledge of the random interval, margin defined by the reliability measure
Monotonic behavior $s \neq 0$	Random interval		Reliability measure	

Table 1.1. *Situations of robustness and reliability for the dimensioning of a resistance*

The schema presented in this section remains static, and needs to be extended to cover the full product design and lifecycles, introducing robustness into design, production, service in normal mode and operations in degraded states in relation to maintenance activities.

1.3. Designing for robust production

In the real world, engineering takes place in an uncertain environment (section 1.1) and the aim of designers is to define products that will be robust and reliable (section 1.2) over the course of their lifecycle. Our focus in this section includes the methodological elements of an approach that aims to create a robust product design process, in order for the production to be robust and for the product to be robust within its environment and conditions of use. After a brief consideration of various development cycle models, we will look at the classic *V* cycle in greater detail, highlighting sources of uncertainty in the model itself and in the course of a typical step in the cycle. The implementation of this approach will be illustrated using diagrams for predicting robustness and mastering uncertainty. The scientific and technical content of this approach will be discussed in later sections in relation to the modeling of uncertain data (section 2.1) and of behaviors (section 2.2).

1.3.1. *Robustness and lifecycles*

The consideration of uncertainty through realistic design [DAN 09] forms part of a global product lifecycle management (PLM) approach, from design to production, which allows permanent validation and virtual certification of the model produced as it will be created and as it will be used[6]. Mastering uncertainty is strategically important when making decisions and implementing actions in the product design and the manufacturing process, in order to reduce the length of the design cycle, optimize industrial procedures and estimate risks, i.e. the consequences of uncertainty, before any decision is made.

To clarify the role of uncertainty in design, we must begin by positioning key concepts in the context of a product development approach. Several models of development cycles are used, including the *V* cycle, the cascade and the spiral (Boehm) cycles [FOR 91, TOL 98, THÉ 03]. These cycles

6 Text based on a contribution by J.-Y. DANTAN, discussed with R. BIGOT and A. ETIENNE.

include all of the phases of verification and validation. The integration of these phases reminds us of the presence of uncertainty: we must verify that specified requirements have been correctly understood, and validate the proposed responses. The spiral cycle places a greater emphasis on risk management than the other two models. The *V* cycle remains the best-known of these models, and will be used as a basis for the analysis presented in this section.

"Designing for robust production" is based on several approaches with slightly different objectives: robust design, reliability-based design, risk-based design and tolerancing:

– The objective of robust design is to identify the solution that minimizes the effect of uncertainty on performance, while guaranteeing required performance levels and respecting requirements.

– The objective of reliability-based design is to demonstrate an acceptable probability level for system failure (i.e. the probability that certain system requirements will not be fulfilled during a given period of time).

– The objective of risk-based design is to identify a solution that minimizes the maximal criticality induced by failures during a given period of time.

– The objective of tolerancing is, for a fixed solution, to determine geometric specifications and allocate tolerances that minimize cost (or another performance criterion), while guaranteeing the required levels of performance and the respect of requirements.

These approaches are not mutually exclusive, and provide complementary views of the effects of different uncertainties. They are included in different phases and stages of the *V* cycle, with a main focus on robust design.

1.3.2. *Description of the V cycle*

The *V* cycle shows the relative positions of the different phases of product development, from specification to validation. In Figure 1.7, the horizontal axis represents time and the vertical axis represents the level of integration: system – parts – components. The descending portion of the *V* cycle includes a functional analysis/specifications and a design phase; the point of the *V* corresponds to the manufacturing and production phase, and the rising part of the cycle shows the different phases of verification and validation. Figure 1.7 (adapted from [AFI 13]) gives a classic view of the *V* cycle, including steps for "client requirements", "responses: physical solutions", etc., and showing a mechanical engineering approach to product development activities: "technical definition of requirements", "concept research", etc. [SCA 04].

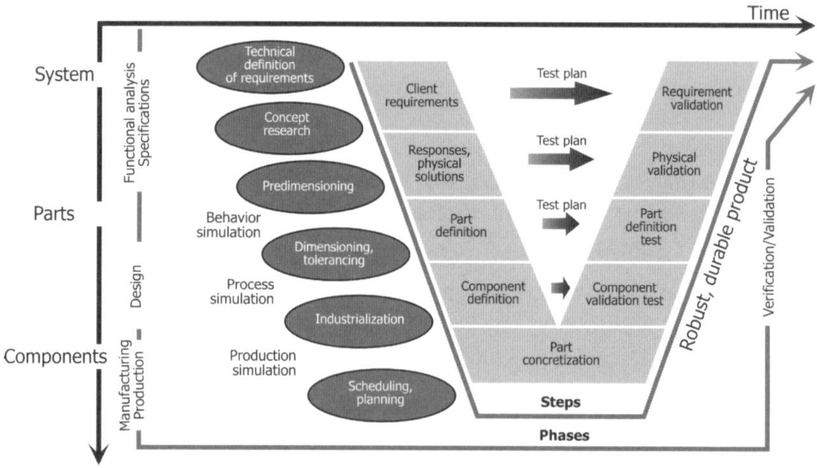

Figure 1.7. *V cycle, completed from a mechanical engineering perspective*

The functional analysis/specifications phase focuses on the definition of requirements, functions, product specifications, etc.; the preparation of a quality plan, validation plan, etc.; and the definition of the desired product reliability level. This phase only covers one step of the *V* cycle: "client requirements", also referred to in mechanical engineering as "technical requirement definition" or "functional analysis" (Figure 1.7).

The design phase then begins by a search for physical solutions (concepts), then the definition of parts (predimensioning), the definition of components (dimensioning, tolerancing, etc.), not forgetting industrialization, sequencing and production planning. This phase runs alongside the definition of test plans and risk analysis activities. The concretization phase (or realization, in the context of systems) is often known as the manufacturing or production phase in the case of manufactured products. The final phase, verification and validation, involves the verification of all components, organs and/or systems in relation to the design, and functional validation – an important phase – in which we consider the functions and quality level of the product in relation to the specification and the requirement analysis:

– Verification: "confirmation through tangible proof that the specified requirements have been satisfied"; verification is a response to the question "are we making the product right?".

– Validation: "confirmation through tangible proof that the requirements for a specific use or planned application have been satisfied"; validation is a response to the question "are we making the right product?".

This brief overview of the *V* cycle introduces the need to consider uncertainty in all of the phases, from specification to validation, particularly for the "definition of the desired product reliability level", "risk analysis" and "verification and validation".

1.3.3. *Uncertainty in the V cycle*

Studies of uncertainty are generally classified according to the origins of the uncertainty:

– Uncertainty relating to the product and its conditions of use (with a direct impact on the performance of the product).

– Uncertainty relating to the overall development process (with an impact on the advancement of the project – uncertainty linked to the project, concerning human operators, activities, planning, etc.).

The analysis given here will concentrate on the first type of uncertainty, although, in reality, the borderline between the two types is extremely flexible. We will now look more closely at the verification and validation phases, highlighting the impact of uncertainty.

1.3.3.1. *The verification phase*

The purpose of the verification phase, for each component of a product or mechanical system, is to detect potential non-conformity to specifications arising from imprecisions inherent in the production process (commonly referred to as manufacturing imperfections) and from uncertainties present in the industrialization phases. The interpretation of component specifications during the industrialization phase can be a cause of non-conformity, but ignorance of certain aspects of the impact of fabrication phenomena on the component is also important. This leads to the classification of uncertainties mentioned in section 1.1.

Uncertainty linked to ambiguity and to the interpretation of data originating in the early phases of the design process (component specifications). The activity of specification interpretation by actors in the production process is not simply a matter of encoding or decoding information, and involves the use of normalized language and "trade" practices. The designer and creator are involved in dual processes of production and interpretation of these specifications, which may be affected by three factors:

– Under-determination or the incomplete nature of component specifications: in this case, it is impossible to reconstruct a meaning from specifications, or this meaning is too general.

– Over-determination, or an excess of specifications: in this case, the interpreter is unable to understand the whole of the specifications, which include several implicit or explicit complementary or contradictory meanings; two types of implicit meanings exist: presuppositions and inferences.

– The use of unclear language, linked to the "vague" nature of certain terms.

Uncertainty linked to earlier phases: manufacturing defaults or imperfections. This is the best-known type of uncertainty. For the purposes of clarification in relation to these cases, we may use an axiom developed by Srinivasan [SRI 99] and Mathieu [MAT 07]:

– *Production variability axiom*: "no-one is capable of producing a component in exact conformity with its functional definition (absolute accuracy); moreover, no one is able to exactly reproduce a pre-existing component (imperfection)".

In other words, it is not possible to produce a component with characteristics identical to those of the target defined by the designer, nor for two components in the same batch to present strictly identical characteristics. This variability, or the manufacturing imperfections, is taken into account in the tolerancing phase with the aim of creating a robust design. At the component level, indicators are used to characterize this variability:

– The dispersion of a characteristic: a bounded interval representing the variability of a characteristic. This value is representative of a percentage of the population (99.73% according to the French standard NFX 06-033).

– Capability[7]: an indicator of the quality of production in relation to a requirement. The required performance, in relation to a parameter – typically a dimension – is characterized by a nominal value $\bar{X}$ and a tolerance interval IT_X. The quality obtained is measured using a sample X_i, $i = 1, ..., n$, with an observed mean of m_X, a standard deviation σ_X and a mean shift $\delta_X = m_X - \bar{X}$. Two levels are defined. The first, denoted as C_p, shows the

7 Reread by N. GAYTON.

dispersion of production and is considered to be equal to $C_p = IT/6\sigma$. The second, denoted as C_{pk}, gives the centering of the production in the tolerance $C_{pk} = \frac{IT_X - 2|\delta_X|}{6\sigma}$. Values of $C_p \geq 1.67$ and $C_{pk} \geq 1.3$ are often used in the automotive industry. Capability levels show both weak dispersion and good centering. Capability levels may be used to deduce the rate of non-conformity, i.e. the number of non-conforming pieces per 10^6 units of production, based on a hypothesis of a distribution law for the sample [GAY 09], generally a Gaussian distribution, which is selected for the simplicity of calculation rather than for objective reasons □.

Uncertainty linked to models, knowledge and processes involved in the activity

– One of the problems encountered during industrialization (involving the modeling of knowledge in order to establish a fabrication process) resides in the fact that certain aspects of this knowledge are not specifically expressed by experts. This lack of precise information results in uncertainty associated with different decisions made during the expert reasoning process [DER 98]. This uncertainty essentially arises from the contextual aspects of the expression of rules. In the case of sequencing machining orders for crossed bores, for example, the general rule is to carry out the smallest boring operation first, but this rule may be supplemented or replaced by different complementary rules: for instance, "if the angle of incidence between the two bores is high and the smallest diameter is large, it is possible to create the larger bore first", or "if the tolerance of the smaller bore is tight, then the smallest bore must be created before the larger bore", etc. The purpose of these rules is to establish connections between different characteristics of the component and the sequencing of operations via a number of contextual criteria. In the first rule, the limit between small and large, in terms of the diameter of the bore, depends on the angle of incidence and remains unclear.

– A second problem is linked to the validation of certain industrialization solutions using numerical simulations of the behavior of the component during shaping; these simulations use models that only constitute partial representations of the real behavior. The granularity of definition of the models used affects the precision of the results, and consequently the understanding and prediction of the phenomena under consideration.

The verification phase, applied at the part, product or system level, has the same essential aims as when it is applied at the component level: its purpose is to detect non-conformity to specifications. This verification is necessary not only due to the imprecisions inherent in manufacturing, but also due to the uncertainties involved in the design phases: concept identification, predimensioning, dimensioning, tolerancing, etc.

The classification of the origins of uncertainty used in these cases is similar to that used in component verification:

– Origins linked to the ambiguity and interpretation of data from previous stages of the design process (product specifications).

– Origins linked to later stages: manufacturing imperfections or imprecisions.

– Origins linked to the models, knowledge and processes used during the various stages of the V cycle, a question covered in some details in sections 2.1 and 2.2.

1.3.3.2. *The validation phase*

The validation phase is different from the verification phase in terms of objectives; the first is functional, whereas the second is of a technical nature. The validation phase serves to ensure that all client requirements have been respected, i.e. to validate the functions expressed in the client specification and described in detail in the functional analysis. This validation is required due to the difficulty of extracting client requirements and transforming them into expected functions, then into technical specifications. The requirement elicitation process involves a variety of communication, negotiation and collaboration activities between future users and the designers of a product, who are most affected by uncertainties linked to ambiguity and the interpretation of data. As little precision is required in general communications, we have a tendency to use imprecise terms and expressions. However, in a context where precision is essential, this lack of clarity can create uncertainty.

This first analysis allows us to position uncertainty as a whole within the product development cycle, and to understand the complexity of mastering uncertainty, due to the wide variety of possible origins. It is important to note that none of the current approaches based on robust design, risk-based design, etc., allow us to tackle uncertainty and its effects in their entirety. The ambiguity inherent in verbal descriptions of processes progressively imposes a rigorous formalization of vocabulary, as in the case of medicine, in order to provide a correct diagnosis, i.e. the existence or non-existence of a pathology.

1.3.4. *Uncertainty linked to a step in the V cycle*

As we have seen, mastering uncertainty is a fundamental and particularly important aspect for all of the activities involved in the *V* cycle. Proactive mastery of uncertainty in the course of these activities requires certain changes in relation to deterministic practices.

It is clearly impossible to identify and model all of the uncertainties involved at the start of the product development process. Certain uncertainties generally decrease over the course of the process, new uncertainties may appear, and a certain number are irreducible. To better understand these uncertainties, we will analyze a generic step in the product development process, illustrated in Figure 1.8.

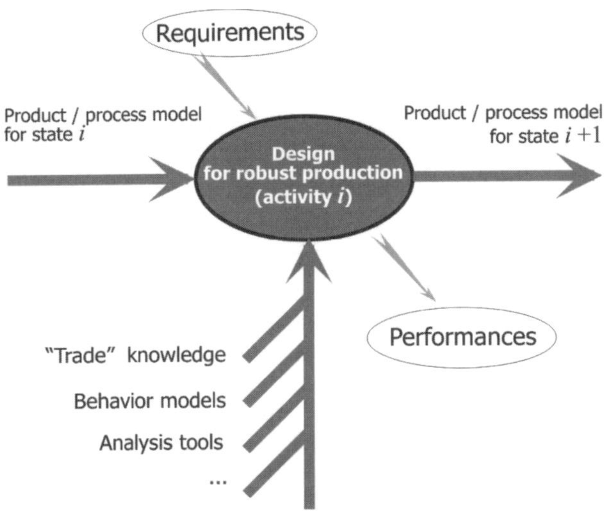

Figure 1.8. *Diagram of a generic step in the product development process*

The aim of a step in this process is to pass from a state i in the definition of the product or the manufacturing process to a state $i+1$, i.e. to obtain additional elements in terms of the definition and modeling of the product or process. This addition is driven by requirements and carried out through the use of domain-specific models, tools or knowledge; it is evaluated using performance measures.

This formalization of a design activity allows us to propose a first taxonomy of uncertainties, using the element connected to the uncertainty as a classification criterion. We can identify a number of different categories.

– Uncertainty *linked to data values*: during a step in designing a product and/or its production process, the generation or evaluation of solutions requires the representation, and therefore the parametering, of the product and/or the process (product/process model of state i or state $i + 1$ – Figure 1.8). These parameters may relate to the geometric description of the product, to materials, to operating conditions of the production process, etc. They are subject to variations in production (imperfections). The variability of these parameters characterizes uncertainty linked to the values of the data used to represent the product or production process during design phases. Figure 1.9 shows the impact of uncertainty linked to a parameter X on performance Y for a minimization of the objective function. See section 2.1 for a discussion of possible representations.

– Uncertainty *linked to the formalization of knowledge*: the difficulty of formalizing the expertise of actors in the product design and manufacturing processes is due to insufficient consideration of context. Expert reasoning and knowledge are based on a set of subjective data (e.g. the choice of the process used to create a hole is a function of not only a set of explicit data, such as the diameter of the hole, but also a set of subjective data, such as the similarity of the hole to other holes, accessibility, etc.). Experts tend to base their reasoning on previous cases, and formalize their knowledge in accordance with these cases. The notions of case similarity and the domain of validity of knowledge are difficult to specify, and pose problems similar to those of granularity in modeling. The limitations of the description of the context of expression result in a non-unified formalization of knowledge. The evaluation of the relevance of a piece of knowledge in relation to a given situation may be characterized by the maturity model. Figure 1.9 illustrates the impact of data on the formalization of knowledge relating to a given objective function, but uncertainty exists in terms of the choice of this function. We cannot be sure if an expert has formalized knowledge in the correct way to allow it to be used in models without generating model errors, i.e. physical mistakes.

– Uncertainty *linked to models* – models or descriptions relating to reality: in the real world, no one is able to create a model or description in exact conformity to reality. A model is, by definition, the partial representation of reality required for a process – the granularity of the definition of the model is chosen in relation to requirements, and cannot tend toward infinity. It is generally the result of a compromise between the limitations imposed on processing resources by high-performance treatments and attaining results with the right level of precision. For this reason, there is a gap between models and reality, which itself is filtered through a model, representing the uncertainty linked to modeling. Work on model shifts, often (incorrectly) referred to as model errors, aims to establish limits or, better, a measure reflecting the uncertainty of a model across a whole domain of definition. These points are discussed in section 2.2.

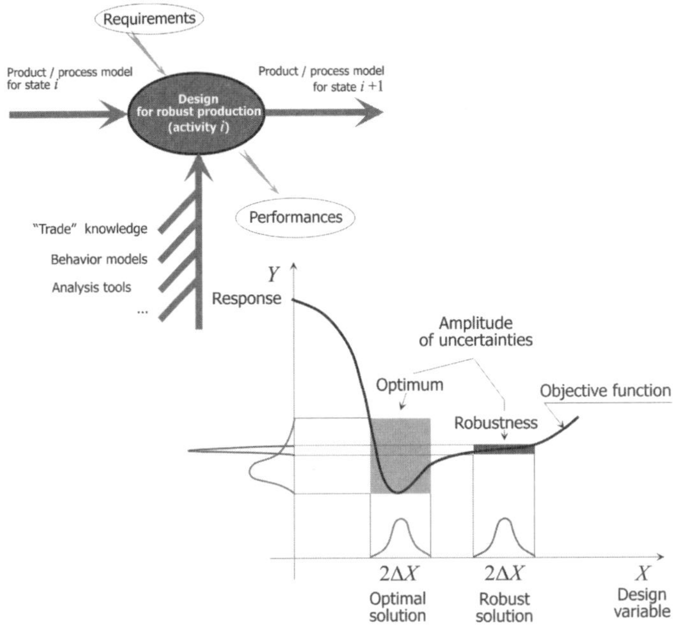

Figure 1.9. *Illustration of uncertainties linked to data values*

– Uncertainty *linked to the modeling of the product and the process* and uncertainty linked to behavior models: for understanding and prediction purposes, "realistic" models and simulations play an essential role in analyzing complex systems such as manufactured products and production process. The granularity of the definition of the models in question (geometric description, description of mechanical behavior, etc.) affects the precision of the results, and thus impacts our understanding and prediction of the studied phenomena. This uncertainty occurs on two levels in relation to design activities: first, activity input/output models, and second, behavior models for specific activities (mostly used in performance evaluation). As we see in Figure 1.10, uncertainty in behavior models, represented by an enveloping zone showing the relationship between the design parameters X and the performance Y on conclusions and decisions that may be made, has a non-negligible impact.

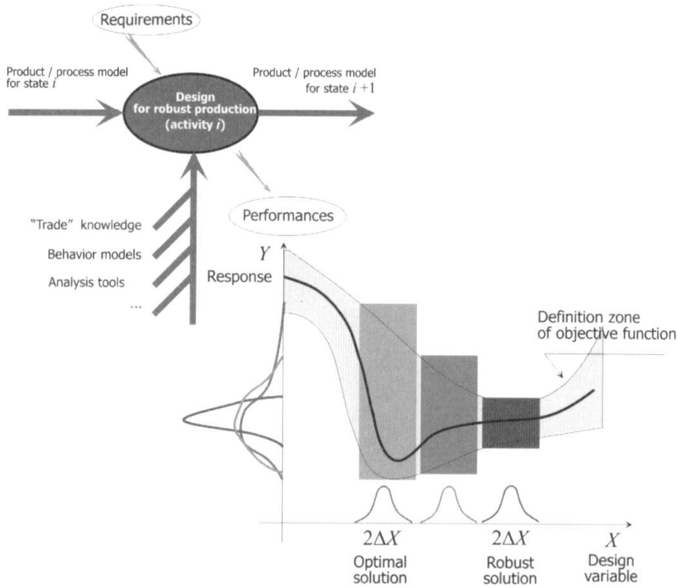

Figure 1.10. *Illustration of uncertainties linked to models and to data values*

– Uncertainty *linked to model couplings*: during multidisciplinary analysis of a product and a production process, the representation of these aspects and the level of granularity used in modeling will not be the same for all experts. The coupling of descriptions involved in multiexpert treatments creates problems in terms of model compatibility; this problem is compounded by limitations linked to the performance of treatment processes (time and precision). This issue is similar to modeling-related uncertainty and to the gap between models and reality; however, in this case, the distance exists between two different models.

– Uncertainty *linked to numerical processing*: in addition to the errors, deviations or simplifications involved in modeling choices, numerical solution is also liable to generate uncertainty, either in "direct" numerical processing of the model (numerical precision, convergence of algorithms) or due to the use of meta-models or response surfaces to avoid difficulties with long calculation times. An approach has been developed in recent years to deal with this type of uncertainty, known as "V&V" for "Model Verification & Validation" [THA 04] □.

1.3.4.1. *Conclusion*

The generation and evaluation of a state $i + 1$ from a state i in order to satisfy design requirements requires an appropriation, formalization and modeling of knowledge. These aspects are driven by the need for predictive performance demonstration. The designer must make use of available information on an as-needed basis. Data interpretation creates knowledge of each aspect, and formalization involves the selection of an appropriate scenario, which is then represented by models. This question is covered in detail in section 2.2. The aggregation aspect of the phase under consideration raises questions concerning performance-related uncertainties; this will be discussed in section 2.3.

In the following sections, we will provide a description of the approach to robustness and to mastering uncertainty used during a V design cycle.

1.3.5. *Robustness and uncertainty*

The aim of the product development process is to create a robust and reliable product[8]. The generic approach proposed in section 1.3.4 may be divided into five actions, categorized by purpose.

1) *Analyze the robustness of a product and the production process in relation to uncertainty (state $i + 1$, Figure 1.10):* analyze the sensitivity of the performance of the "manufactured" product to uncertainty (as a whole) or to certain specific uncertainties. Measure the difference between expected performances and those exhibited by the "manufactured" product.

Example: analyze the sensitivity of the operational strength of a part to variations in material characteristics inherent to the manufacturing process.

2) *Analyze the robustness of a solution to calculations in relation to calculation models*: analyze the sensitivity of calculated results to uncertainty linked to data values, uncertainty linked to descriptions of the product, process and behavior, and/or uncertainty linked to the treatment.

Example: analyze the sensitivity of the results of a simulation to differences between reality and the behavior model.

3) *Reduce the sensitivity of the performances of the "manufactured" product*: in relation to uncertainty (as a whole) or to certain specific types of uncertainty. Reduce the gap between the expected and real performances of

8 Text based on a contribution by T. YALAMAS.

a manufactured product by modifying the solution (product and/or production process and/or processing tool and/or granularity of modeling, etc.).

Example: improve the behavior model in order to reduce the sensitivity of simulation results to differences in relation to reality.

4) *Optimize the solution(s) (product and/or production process and/or processing tool and/or granularity of modeling, etc.)*: to minimize the sensitivity to uncertainty of the performances of the manufactured product, or to guarantee a certain level of sensitivity to uncertainty in the functions or characteristics of the manufactured product.

Example: optimize the targets of "product" parameters in order to minimize the sensitivity of product performances to parameter variations.

5) *Tolerance*: limit overall or specific uncertainties in order to guarantee a probable performance level for the solution (product and/or production process and/or processing tool and/or granularity of modeling, etc.).

Example: limit variations in material characteristics in order to guarantee a certain level of operational strength or allocate reliability across a system.

Each of these actions may be the target of specific developments, and we will discuss certain aspects in greater detail in relation to sensitivity and reliability analysis (section 2.2) and multicriterion optimization (section 3.1). However, robustness and reliability cannot be isolated within specific parts of the development cycle, but must be integrated into a global approach based on two main components: robustness and uncertainty.

1.3.5.1. *The robustness approach*

Figure 1.11 shows mastering robustness in a design process as a process in its own right, split into different steps corresponding to phases of the design process.

This illustration demonstrates the complementarity of approaches, often carried out by different entities within a group, which all aim to ensure that the system responds to a set of requirements. Note the importance of strong communication links between services to ensure that all of the available information is used in the best possible way.

In the downward section of the *V* cycle, classic tools such as functional analysis, preliminary risk analysis and failure mode, effects and criticality analysis (FMECA) are used; these are relatively common [COL 11] and do not require detailed treatment here. The use of simulation for mastering uncertainty is somewhat less common, and will be discussed in greater detail below.

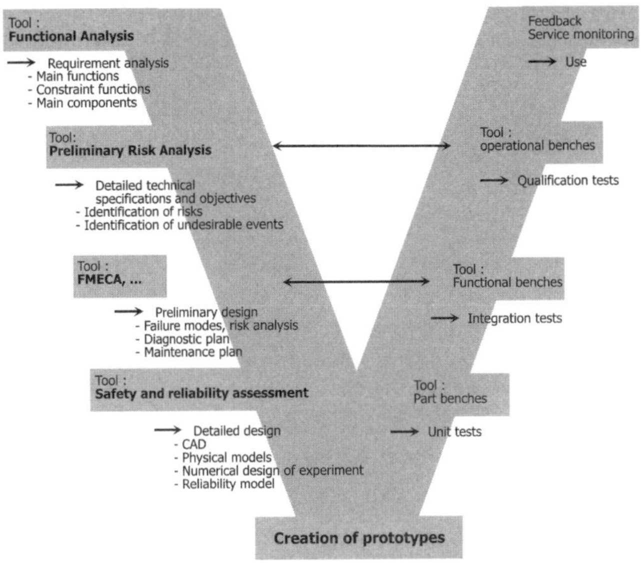

Figure 1.11. *Robust and reliable design approach using the V cycle*

1.3.5.2. *The uncertainty approach*

Much work has recently been carried out on modeling uncertainty and its propagation through a physical (mechanical, hydraulic, chemical, etc.) model, involving discussions between actors from a number of different establishments (notably *Electricité de France* (EDF), European Aeronautic Defence and Space Company (EADS), Phimeca Engineering S.A (PHIMECA) and the Commissariat à l'Energie Atomique (CEA) in France). These discussions resulted in the publication of a book on the subject, [DE 08]. This methodology, taken in a probabilistic context, consists of four clearly identified steps (Figure 1.12).

– *Step A – Physical model*

From an engineering perspective, this is the most widely known step. It consists of the following:

- Identifying the system in question and modeling it from a deterministic standpoint (analytical model, finite element model, etc.). In cases where detailed modeling would generate excessive calculation costs, it is also possible to replace the deterministic model at this step by a predefined meta-model (or response surface) obtained using polynomial chaos [SUD 07b], support vector regression (SVR) [DEH 08b] or by Kriging [DUB 11b, ECH 12]), or to use model reduction techniques (see section 2.3).

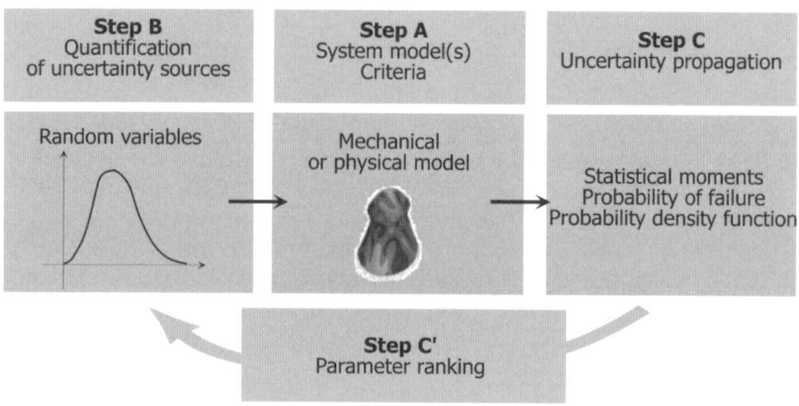

Figure 1.12. *Methodological principles for the treatment of uncertainty, based on [DE 08]*

- Identifying the model output values that are of interest (maximal constraints in a domain, constraints at a point, displacement components, etc.).

- Formulating a criterion using one or more quantities of interest, where necessary (e.g. for reliability analysis).

– *Step B – Quantification of uncertainty*

At this step, we distinguish between cases based on the availability or unavailability of data required to construct a probabilistic model (see section 2.1).

- In cases where experimental data or a database is available, we may use the usual statistical methods to identify the most suitable random variable distribution.

- In the absence of data, *a priori* physical knowledge of the parameter being modeled should be combined with the experience of the probabilistic engineer in order to construct a physically acceptable probabilistic model.

Note that model updating methods [BER 11] allow the use of measures obtained *a posteriori* to refine an existing probabilistic model in order to improve the predictive capacities of the model.

– *Step C – Propagation of uncertainty*

Three types of analysis are generally used:

- Scatter analysis (Figure 1.13(a)): its aim is to characterize (in the statistical sense of the term) one or more quantities of interest. It involves the determination of the statistical moments (e.g. mean and standard deviation) of this quantity of interest, up to an order that depends on the desired precision and on the chosen method.

- Distribution analysis: (1.13(b)): in this case, the aim is to estimate the complete probability distribution of the quantities of interest.

- Reliability analysis (Figure 1.13(c)): the purpose of reliability analysis is to obtain as precisely as possible the probability that the parameter in question will exceed a threshold value: the probability of failure in relation to the dimensioning criterion [LEM 09].

Note that implementation difficulties increase as the distance from the central trend (Mediocristan) increases, and that we wish to estimate the tails of distributions (Extremistan) (from left to right in Figure 1.13).

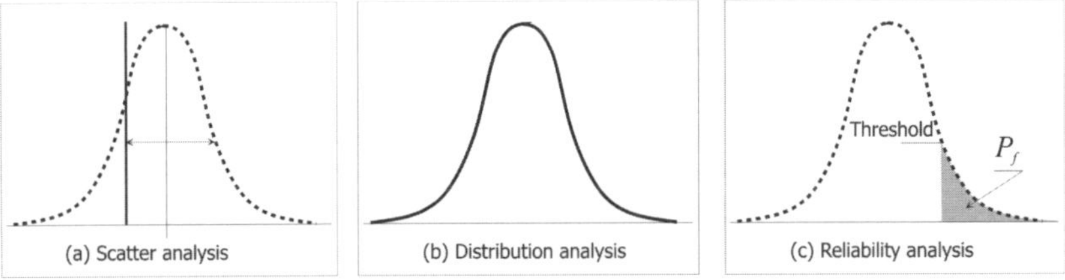

Figure 1.13. *Step C: Propagation of uncertainty*

– *Step C' – Ranking uncertainties*

- The purpose of this step is to identify parameters, or combinations of parameters, where variability has the greatest impact on the quantity of interest. This step is generally associated with the previous step as the methods used in step C also provide information for ranking. A distinction is made between local sensitivity indexes, generally associated with reliability analysis, and global sensitivity indexes, which are often associated with a variance breakdown analysis known as analysis of variance (ANOVA) [SAL 04,

SAL 10]. Figure 1.14 shows a classic representation of the weightings of four variables □.

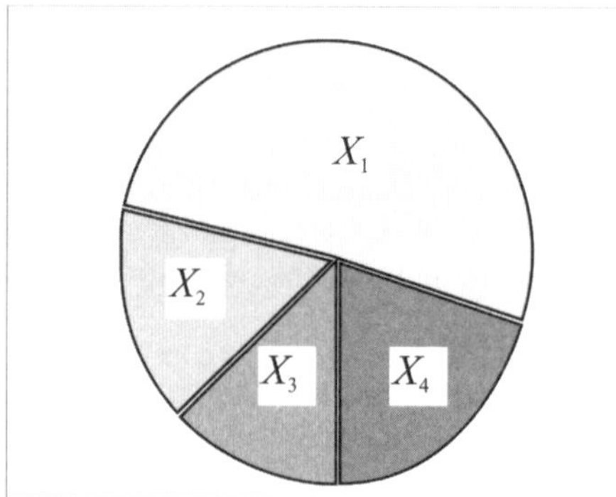

Figure 1.14. *Step C': Illustration of uncertainty weightings for variables: X_1 to X_4*

1.3.6. *Conclusion*

In designing products, parts or systems for robust production, we work in an environment of dynamic and varied uncertainty, which comes from a wide range of sources. These sources include ambiguity in projects, which are often described in insufficient detail; knowledge, which must always be limited; and uncertainty related to the use that the user will make of the product. This uncertainty is dynamic, as it evolves as a function of decisions and the advancement of knowledge. Designing for robust production requires us to integrate all of the necessary phases and steps, from those based on cognitive sciences to those involving different branches of physics, mechanics and mathematical modeling. Individual aspects cannot usefully be considered in isolation. For this reason, designing for robust production requires the involvement of multidisciplinary teams, with each member contributing fully to each step and phase in the process.

The *V* cycle and the associated approaches for demonstrating robustness and reliability provide bases for reflection, which is now at a stage requiring better formalization. The Wikipedia article on the subject cites [KEY 78], which uses a hexagram diagram, Figure 1.15 (left), to represent the interdisciplinary links between the cognitive sciences. Similarly, we can create a hexagram of the links between modeling sciences, shown in

Figure 1.15 (right). The algebra of the operators connecting these entities needs to be constructed in order for design for robust production to become a science in its own right, providing the elements required for decision acceptability from all relevant perspectives: societal, economic, judicial, technical, etc., as was seen at the IMdR/AFM conference in May 2012 [LEM 12].

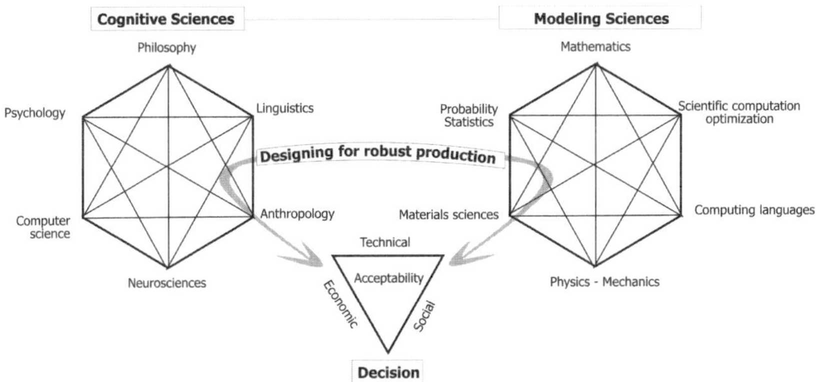

Figure 1.15. *Symbolic representation of the cognitive sciences (left) and the modeling sciences (right). Designing for robust production requires sustained and effective dialog between the two domains in order to produce acceptable decisions*

In the following chapter, we will consider a number of points relating to the field of modeling.

Chapter 2

Modeling Uncertainty

2.1. Random uncertainty

2.1.1. *Modeling uncertainty*

In this section, we will consider the representation of random uncertainty. From a formal perspective, a quantity of interest is represented by a set of functions of space, with variable x belonging to a domain D; time, with variable t belonging to a lifetime T; and uncertainty, with variable ω belonging to a certain probability space Ω. Let $X_i(x, t, \omega)$ be the functions for which the space Ω is the result of the definition subspaces Ω_i of each, integrating the internal dependencies on space and time (autocorrelation) and dependencies between functions (intercorrelation). The impossibility of exploring a space of this type in its entirety means that considerable simplifications are required.

From the simplest: as a general rule, in engineering, uncertainty is expressed using two quantities: a characteristic value x_k, representing an appraised value such as the mean or median, raised or lowered depending on which direction is most favorable in terms of safety and a safety factor γ_{X_k} taking account of event-based feedback. The definition of these quantities was initially based entirely on expertise and experimental feedback, which was then codified. Nowadays, probabilistic methods allow for better exploration of the structure of uncertainty using calibration methods [CEN 03, LEM 09], and expert opinions are used to justify the probabilistic model. Note that the pairing (x_k, γ_{X_k}) is not intrinsic to a variable, but is dependent on the behavior model in which it is included. The possibility of dependency is taken into account by combination values. This approach has generally proved effective in creating a margin between the median point of operation and the border of Extremistan; however, it gives neither the measure nor the structure of this margin, limiting the possibilities for optimization, particularly when attempting to demonstrate the existence of an acceptable cost to failure risk-balance.

To the most complex: probability theory constitutes a perfect mathematical framework, but one based on information, which is never sufficient to describe the space Ω. A calculated probability is simply a value conditioned by the hypotheses used in constructing the stochastic data model; an exploration of Mediocristan will allow us to define these hypotheses in a better way.

2.1.2. *Exploration of Mediocristan*

2.1.2.1. *Using data that are really available*

In exploring Mediocristan, we reduce epistemic uncertainty to obtain a mapping, which is sufficiently precise for our needs. Knowledge of a variable may take different forms:

– simple expert opinions regarding the order of magnitude of a mean, median or mode;

– the definition of an interval defining the scale of variability;

– values of fractiles at different levels;

– the first statistical moments;

– all of the statistical moments: this is an ideal case and is equivalent to knowledge of the probability distribution.

To this, we must add physical constraints, such as a positive characteristic (the thickness of sheet metal, for example, can never be negative). The passage to a probabilistic model can introduce unproven information, if the necessary precautions are not followed. A number of possible probability distributions exist based on the first two moments (mean and variance), and it is not possible to choose between these distributions without using other elements. Two types of reasoning are generally used in this choice. The central limit theorem directs us toward a Gaussian distribution, which is not physically admissible, and the maximum entropy principle (characterizing the broadest probabilistic information) pushes us toward a uniform distribution using a support $[a, b]$. This raises two issues: first, we need to determine whether or not the bounds a and b are certain, and second, the attribution of a uniform density, as suggested by the principle of maximum entropy, involves the unreasoned attribution of a structure. If the expert defines an interval $[a, b]$ and has an idea of the mean m, the only case that does not generate a contradiction will be $m = (a + b)/2$.

These remarks highlight the need for extreme prudence when moving from raw information to a data model.

2.1.2.2. *Quantitative approaches*

[1]Before we examine the probabilistic model in section 2.1.3, we will briefly consider quantitative approaches, which are less demanding in terms of available information. This question has been addressed in some depth by D. Dubois and H. Prade. A summary is given in [BAU 06a].

2.1.2.2.1. Fuzzy sets

The theory of fuzzy sets was introduced by Zadeh as an extension to the classic theory of sets [ZAD 65]. According to set theory, membership of a set is binary: either an element is part of a given set or an element is not part of the set. The theory of fuzzy sets allows partial adherence, meaning that the membership of an element is represented by any real value taken from the closed set $[0, 1]$. According to conventional Boolean logic, a statement is true or false, and the proposition "element B is a member of set $\mathcal{F}$" has a truth value of 0 or 1. In fuzzy logic, the statement can take a fuzzy truth value (measure) given by any real number in the interval $[0, 1]$. If we specify, for example, that proposition "element B is a member of set $\mathcal{F}$" has a fuzzy truth value of 0.3, this means that element B is only a partial member of set $\mathcal{F}$. This means that set $\mathcal{F}$ is fuzzy, as the limit of membership is not clearly defined. Fuzzy logic has been used extensively in the form of fuzzy controllers, a generalization of expert controllers [LEE 90]. Effective applications have also been developed in linguistics, where the meaning of a word can often remain vague or ambiguous, with a degree of acceptance that varies according to the context [MER 93]. If a man is described as tall, for example, how tall is he? What is the minimum height required to be considered tall? This question has no universally acceptable response, as the answer depends on personal interpretation. Fuzzy logic has also been applied in photography in calculation algorithms for optimal speed, aperture and International Organization for Standardization (ISO) settings.

1 Text discussed with S. CASTRIC. Reread by F. AGUIRRE.

While fuzzy logic has proved valuable in both control and linguistic applications, its application within the context of mechanics remains limited.

2.1.2.2.2. Possibility theory

Possibility theory, an extension of fuzzy logic, was proposed by Zadeh [ZAD 78] as a tool for representing information expressed in terms of fuzzy measures. Possibility theory defines two evaluations [DUB 00] showing the membership of a variable x within an interval A:

– the possibility: $\Pi(A) = \sup_{x \in A} \pi(x)$;

– the necessity: $\mathrm{N}(A) = 1 - \Pi(\overline{A}) = \inf_{x \notin A}(1 - \pi(x))$.

Any pair of functions $[\Pi(A), \mathrm{N}(A)]$ may be interpreted as the upper and lower bounds given by specific probability families. There is a clear relationship between probabilistic approaches and possibility theory. The use of possibility theory is justified in cases where little information is available; probability theory is preferred when sufficient data are available to produce a stochastic model.

One argument put forward in favor of the use of possibility theory in design is the simplicity of the operations involved (for example see Du and Choi [DU 04]). These operations are concise and rapid, with no joint distributions or other complex relationships; however, it may be seen as an excessive reduction of the physical reality.

2.1.2.2.3. Theory of belief functions

Belief functions constitute the framework for the theories developed by Shafer [SHA 76], based on ideas taken from Hooper, Bernoulli and Lambert [SHA 78, SHA 86], in extending work carried out by Dempster [DEM 67]. The term "Dempster–Shafer theory" is also used.

Let us take a sample of $i = 1, ..., K$ measurements m_i subject to imprecision due to a perception threshold δ. The sample is therefore composed of K random intervals of the form $[m_i - \delta, m_i + \delta]_{i=1,...,K}$, each associated with a probability ν_i of observing the measured value m_i. If $\delta = 0$, this is the probability of m_i; otherwise, the probability is that of the interval. Measurements m_i form an exclusive and exhaustive set of K results, known as the frame of discernment Θ, whose elements are focal elements E. The approach is similar to that used in probabilities, but differs in the use of subsets of the frame of discernment [BAU 06a].

Let us consider the case $K = 3$. In this case, $\Theta = \{m_1, m_2, m_3\}$ and the full list of subsets is $\{m_1\}$, $\{m_2\}$, $\{m_3\}$, $\{m_1, m_2\}$, $\{m_1, m_3\}$, $\{m_2, m_3\}$, $\{m_1, m_2, m_3\}$. According to the available data, each of these subsets is accepted with a certain degree of belief. There may be evidence, for example, which supports $\{m_1\}$ and $\{m_2\}$ while excluding $\{m_3\}$, or which suggests that no distinction should be made between $\{m_1\}$ and $\{m_2\}$.

Two functions qualify the proposition that a variable X is a member of an interval A:

– the belief, defined by $\mathrm{Bel}(A) = \sum_{E,\, E \subseteq A} \nu(E)$;

– the plausibility, defined by $\mathrm{Pl}(A) = \sum_{E,\, E \cap A \neq \phi} \nu(E) = 1 - \mathrm{Bel}(\overline{A})$.

Belief functions form a framework for possibilistic and probabilistic theories, and constitute a specific instance of imprecise probability theory. Aguirre–Martinez's doctoral thesis [AGU 12] shows the importance of belief functions in demonstrating system reliability.

2.1.2.2.4. Probability box (p-box) [FER 03]

Let us suppose that we know the mean m and interval $I = [a, b]$ of a variable with an unknown distribution function $F(x)$. Distributions $\bar{F}(x) = \sup_{F:\mathrm{E}[X]=m} F(x)$ and $\underline{F}(x) = \inf_{F:\mathrm{E}[X]=m} F(x)$ contain a family of probabilities. If the variable is known based on a sample, the Kolmogorov–Smirnov test may be used to construct a confidence interval (but no bounds) based on the empirical distribution function $\square$.

2.1.2.2.5. Conclusion

This overview highlights the importance of not introducing more information than is really available into a model. The tools discussed above are still undergoing development, and should be valuable in improving probabilistic modeling, often the only option when modeling dependent variables which are functions of time and space and in propagating uncertain variables in behavior models.

2.1.2.3. *Robustness and convexity [HAI 95, HAI 90]*

Let us consider a known realization $\{\bar{u}\}$ of the definition space of the variable model and a real number α. The set of possible realizations $\{u\}$ situated at a distance α is given by:

$$\mathcal{U}(\alpha, \{\bar{u}\}) = \left\{ \{u\} \mid \|\{u\} - \{\bar{u}\}\|^2 < \alpha^2 \right\}$$

for a chosen norm $\|.\|$. This definition generalizes the notion of a robust interval to the convex function of dimension α, a parameter of uncertainty in relation to the central term $\{\bar{u}\}$. The robustness (according to the definition used by the authors) is then associated with the highest possible value of α.

Let $\{u\}$ be the vector of the input parameters of a system and $\{x(\{u\})\}$ be the response given by a model $\mathcal{M}$: $\{u\} \rightarrow \{x\} = \{x(\{u\})\} = \mathcal{M}(\{u\})$. The set of responses:

$$\mathcal{X}(\mathcal{U}) = \{x(\{u\}) \,|\, \{u\}\} \in \mathcal{U}$$

may be classified, as a function of what is considered acceptable, into a success set and a complementary failure set $\mathcal{F}$. The convexity hypothesis, which is verified if the model is linear, allows us to seek the widest possible dilation of the convex function such that sets $\mathcal{X}$ and $\mathcal{F}$ are tangents. The dimension of the convex function shows the quantity of uncertainty that the system is able to support safely, i.e. in a way which is acceptable for the decision maker. A probabilistic model would require an acceptable probability value; the convex model requires us to define the dimension of the convex function, i.e. the acceptability of minimal performances.

This approach has been applied in the domain of reliability and for the validation of models [HAI 04].

2.1.2.4. *Conclusion*

In exploring Mediocristan, we assemble the relevant available information in order to obtain a direct probabilistic description or to correct a previous *a priori* description. While the proposed methods are relatively easy to apply when modeling a variable, only probability theory currently produces a modeling, which is compatible with the representation of stochastic processes or random fields and which may be implemented in behavior models. Note the presence of an essential scientific barrier in this case: the choice of the form of information used to represent variability based on the available information.

In the following section, we will discuss certain elements of statistical models used in the construction of stochastic models.

2.1.3. *From statistics to probabilities*

2.1.3.1. *Parametric estimation*

Descriptive statistics uses a data sample to construct a histogram and calculate estimators. Mediocristan is characterized not only by the robustness of statistical estimators – the mean, the median and the variance – but also by the physical considerations, such as the obligation to use positive values, or bounds. Note that these estimators simply apply to a sample, and are not parameters of the population. Statistical theory may be used to calculate confidence intervals for estimators, but mathematicians should be aware of the strict hypotheses involved, which do not allow delimitation of a variance without the Gaussian hypothesis. In mechanics, we must adapt to these hypotheses. Estimators are therefore random variables for which we require estimators, and so on. The territory of Mediocristan is therefore fractal, and only the first modes are accessible.

A large number of probability distribution functions (PDFs) are possible *a priori* in cases where only the first statistical moments are available (generally the mean and the variance) and tests (χ^2, Kolmogorov–Smirnov, Anderson–Darling) may be used in order to not reject (but not to accept) a hypothesis. Criteria such as the Akaike information criterion (AIC) or Bayesian information criterion (BIC) use a likelihood function to select particular models. The limitation of the parametric estimation is that the choice of a PDF introduces wrong information and physically impossible realizations can appear. For this reason, statistical optimization must be carried out under the constraint of physical admissibility, for example positive values and a bounded support, meaning that the Gaussian law is not appropriate. We must also take account of the role of the variable in a behavior model. Thus, if a variable appears as a denominator, it cannot include potentially null realizations. A notable exception to the approach described above is the Weibull distribution. Unlike the statistical approach, the brittle fracture mechanism of a material from defects of random size is used to justify the expression of the law.

2.1.3.2. *Non-parametric estimation*

Rather than trying to "fit" a probability distribution, non-parametric estimation seeks a function $f_X(x), x \in D$ where the only required property is $\int_D f_X(x)\, dx = 1$. The histogram is smoothed by introducing a kernel, i.e. a function for each class in the histogram; these kernels may be triangular, Guassian, Epanechnikov, etc. This procedure does not introduce *a priori* assumptions concerning the distribution, but is dependent on the choice of a kernel.

When attempting to avoid *a priori* choices, it is also possible to develop any random variable using polynomial chaos. This notion, which is essential both in terms of the identification and the propagation of uncertainty, will be developed in more detail in section 2.1.4, which also illustrates the passage from representing a variable to representing a field or a process.

2.1.3.3. *Independent and dependent variables*

The discussion in the previous sections was based on the estimation of a variable. However, data models often concern variables for which the uncertain characteristics are linked. The notions of independence and dependence should be addressed as a function of the stochastic model. From a probabilistic perspective, variables are independent if their joint density is the product of the marginal densities. A hypothesis of this kind may often be formulated for physical reasons (for example the independence of properties of materials and the static actions applied to them). In cases where statistical dependence is suspected, the copula samples are analyzed in order to determine the correlation between variables; note, however, that a correlation of zero does not imply independence except in the case of Gaussian variables. The usual measures used are Pearson's linear correlation, Spearman's correlation or Kendall's rank correlation. These involve hypotheses that are not always justified and only consider second-order mixed moments. The most general formulation is that of the copula relating to a vector of random variables. This formulation describes the function that connects the set of marginal distribution functions to the joint distribution of the vector. A number of models exist, which then require adjustment based on observations; this is not an easy process. Sklar's theorem states that if C is a copula, and if $F_1, F_2, ..., F_d$ are (univariate) distribution functions, then $F(x_1, ..., x_d) = C(F_1(x_1), ..., F_d(x_d))$ is a distribution function of dimension d, for which the margins are precisely $F_1, F_2, ..., F_d$, and reciprocally, if F is a distribution function of dimension d, a copula C exists that is given by the relationship $C(u_1, ..., u_d) = F\left(F^{-1}(u_1), ..., F_d^{-1}(u_d)\right)$. The copula of the random vector $(X_1, ..., X_d)$ is therefore the distribution function of the random vector $F_1(X_1), ..., F_d(X_d)$. The interest of the copula lies in the fact that it contains all of the connections resulting from the statistical moments of the combined variables; these moments are null in the case of independence. The specific case of the linear correlation of Gaussian variables is known as the Gaussian copula. The concept of copulas is widely used in the financial sector (see [KHA 08] or [NEL 99] for further details).

Copulas are also used in mechanics [LEB 08, DUT 09, LEB 13c], in probabilistic safety evaluations and in engineering problems

[ACH 10, ACH 12]. When considering reliability, the choice of a copula to represent dependencies in distribution tails is an important issue to consider.

Bayesian networks constitute another pathway for exploration and will be discussed in more detail in section 2.2.2.5.2.

2.1.3.4. *Objectivity and subjectivity*

Section 2.1.3.1 is based on objective observations, which, using probability theory, results in the use of a frequency approach. The Bayesian approach [DRO 02, BOR 10], known as the subjective approach, also offers responses to modeling questions. Instead of processing the available information and looking for the best probability distribution, this approach is based on *a priori* knowledge of the probability distribution, which is then updated using new information. Bayesian inference consists of using expert knowledge to choose a density distribution $f_X(x, \theta)$, where θ is the parameter vector, and an *a priori* distribution for each parameter $p_\Theta(\theta)$. The *a posteriori* parameter distribution is obtained using:

$$f_\Theta(\theta) = \frac{1}{c}\, p_\Theta(\theta)\, L(\theta, x_1, ..., x_N)$$

where c is a normalization constant and $L(\theta, x_1, ..., x_N)$ is the likelihood function of the observed sample, equal to $\prod_{i=1}^{N} f_X(x_i, \theta)$ for independent observations. This procedure illustrates the combination of treatments of aleatoric uncertainty by the choice of a distribution and epistemic uncertainty through learning parameters. It should also be possible to update the distribution, for example by using a family of distributions.

2.1.3.5. *Robustness of the estimation*

A sample is never exhaustive, and the quality of an estimation depends on the size of the sample, N. A confidence interval is a function of $\sqrt{N}$, which shows weak convergence. The robustness of the estimation is measured by resampling techniques. These techniques consist of constructing a statistic based on subsamples obtained by removing one or more elements, with or without replacement. These bootstrap, jackknife or leave-one-out techniques [EFR 79] are used to obtain confidence intervals.

2.1.3.6. *Conclusion*

The passage from observations and samples to a probability distribution is a difficult and sensitive exercise, based on both mathematical and physical coherency: modeling should not introduce events that are artificial or of a non-probabilistic nature, and information is always limited in terms of space and time. In [ZEY 13][2], the authors underline the fundamental conditions to build an experimental probabilistic hypersurface (EPH) according to the probability theory: this requires physical independence, no *a priori* regarding predefined distributions, adaptiveness to varied information, transparency for users, mathematical rigor and robustness.

However, it is essential to ensure knowledge of the physical nature of the subjacent model, which is itself subject to uncertainty, something we will examine in section 2.2. Without this knowledge, we are subject to the same fate as N.-N. Taleb's turkey: each day for a year, a turkey notes its level of satisfaction. The day before Thanksgiving (or Christmas), the turkey feels particularly well cared for, and predicts a glorious future with a high degree of probability.

2.1.4. *Polynomial chaos*

[3]Polynomial chaos expansion is a probabilistic method that may be used to represent an uncertain quantity of interest (a variable, process or random field) based either on a parameterized distribution, which introduces a powerful tool for propagating uncertainty (see section 2.3.2.4), or a data sampling, with no absolute requirement for *a priori* knowledge of the distribution of this quantity of interest.

In general terms, the polynomial chaos expansion formalizes a separation between stochastic and deterministic components of the quantity of interest. Polynomial chaos is a spectrum representation based on the projection of random values into a probabilized space Ω. The deterministic part is modeled by stochastic modes, weighting random polynomial functions; this separation

---

2 The title "Probabilistic Information Transfer" refers to statistical inference, and not to the propagation of uncertainty as described in section 2.3.

3 Contribution by S. BERGER.

is analogous with that used in modal dynamics. Let $\omega \in \Omega$ be the random character of a variable and x its indexing in space or time. We obtain:

$$X(x,\omega) = \sum_{j=0}^{\infty} \phi_j(x)\,\Psi_j(\omega)$$

where $\phi_j(x)$ are the modes and $\Psi_j(\omega)$ is a base of random polynomial functions.

Several types of polynomial chaos expansion exist, depending on the choice of an orthogonal basis for the projection of the random variables. The earliest and best-known form is Wiener–Hermite polynomial chaos, based on Wiener's theory of homogeneous chaos, and the Cammeron–Martin theorem, according to which a second-order stochastic process may be expressed by a convergent expansion (in the sense of least squares) as a series of Hermite polynomials, orthogonal functions of independent Gaussian random variables [WIE 38, CAM 47, GHA 91].

As the speed of convergence, and consequently the precision of the expansion in Hermite polynomials, is not verified in the case of non-Gaussian processes, the idea has been generalized to non-Gaussian random processes to define a correspondence, known as the Askey scheme, between families of orthogonal polynomials and probability distributions [ASK 85]. The notion of generalized polynomial chaos, or Wiener–Askey polynomial chaos, is thus defined and an exponential convergence is demonstrated and generalized to cases using different probability distributions [XIU 02, XIU 03]. It thus becomes possible to represent any form of chance contained in a distribution function or in a sample [SOI 04].

In practice, expansions are truncated at a certain order. Thus, the construction of a chaos expansion requires us to calculate a finite number of stochastic modes using non-intrusive methods, i.e. methods based on legacy computer software with no modifications. These only require a limited number of values for the quantity of interest, or, in other terms, a limited sampling of numerical experiences. Two types of methods are proposed: regression methods and spectral projection methods. In regression methods, the stochastic modes are calculated by minimizing (in the least squares sense) the difference between the random quantity of interest and its approximation in the polynomial chaos base [CHO 04]. Using non-intrusive integral projection methods, the stochastic modes are determined by the projection of the random function into the chaos base; the stochastic modes are then obtained using numerical calculations of integrals [CRE 09]. The best-known decomposition is the Karhunen–Loève decomposition [KAR 46, LOE 55], also known as the proper orthogonal decomposition (POD). It is based on the eigenvalues of the autocorrelation function of the process.

Using stochastic modes, the polynomial chaos expansion provides us with an estimation of the random quantity of interest, allowing us to determine statistical moments up to a chosen order and to reconstruct the density to a level of precision that depends on the order of truncation.

One of the main advantages of chaos expansions is that they do not require advance knowledge of the distribution of the quantity of interest. Nevertheless, reservations may be imposed due to the fact that the speed of convergence of the chaos expansion is dependent on the optimal choice of a chaos base, which is linked to the type of distribution of the variable of interest. A second advantage is that chaos only requires a limited number of values for the quantity of interest; however, the higher the number, the higher the order of truncation and, consequently, the precision of the expansion.

The limitations of chaos are essentially due to the inevitable truncation, which limits us to an approximation of the quantity of interest. Thus, convergence problems, with the difficult choice of the number of expansion terms, are linked to the non-existence of a mathematical criterion allowing error control and, finally, validation issues remain relevant □.

2.1.5. *Exploration of Extremistan*

The exploration of Mediocristan essentially involves interpolation, while the exploration of Extremistan constitutes a hazardous extrapolation to events that have, in some cases, been observed, but in other cases are merely presumed to have happened, while feedback demonstrates the existence of numerous successes. This extrapolation is always the result of incomplete data, as the sample is obscured at the last point of observation.

The exploration is based first on an extension of knowledge of Mediocristan, and then on the application of principles linked to an acceptability decision.

2.1.5.1. *Extreme value statistics*

Extreme value statistics concerns the tail of the distribution of a variable, and thus supposes a continuity from Mediocristan into Extremistan. Thus, the maximum of a random observation repeated over time (or space) allows a generalized extreme value (GEV) distribution function [RES 78]:

$$G(x,\mu,\sigma,\xi) = \text{Prob}\{X \leq x\} = \exp\left\{-\left[1+\xi\left(\frac{x-\mu}{\sigma}\right)\right]^{-1/\xi}\right\}$$

with $1+\dfrac{\xi\,(x-\mu)}{\sigma} > 0\,;\mu \in \mathbb{R}\,;\xi \in \mathbb{R}\,;\sigma > 0$

where ξ is the shape parameter, μ is the position parameter and σ is the scale parameter. This function encompasses the classic Gumbel, Fréchet and Weibull (extreme) distributions:

$$\text{Gumbel} : \xi \to 0 \Rightarrow \exp\left(-\exp\left(-\frac{x-\mu}{\sigma}\right)\right) \quad x \in \mathbb{R}$$

$$\text{Fréchet} : \xi > 0 \Rightarrow \begin{cases} 0 & \text{if} \quad x \leq \mu \\ \exp\left(-\left(\frac{x-\mu}{\sigma}\right)^{-1/\xi}\right) & \text{if} \quad x > \mu \end{cases}$$

$$\text{Weibull} : \xi < 0 \Rightarrow \begin{cases} \exp\left(-\left(-\frac{x-\mu}{\sigma}\right)\right)^{-1/\xi} & \text{if} \quad x < \mu \\ 1 & \text{if} \quad x \geq \mu \end{cases}$$

The peaks over threshold (POT) approach is often used in engineering to define extreme values (in seismology, hydrology, etc.). The distribution function of excesses $X - u$ above the threshold u conditional on $X > u$ becomes:

$$F_u(x) = \text{Prob}\left\{X \leq u+x\,|\,X > u\right\} = 1 - \frac{1-F(u+x)}{1-F(u)} \qquad x > 0$$

The Pickands theorem [PIC 75] shows that the function $F_u(x)$ must tend toward the generalized Pareto distribution (GPD):

$$P_{\xi,\mu,\sigma}(x) = \begin{cases} 1-\left(1+\xi\frac{x-\mu}{\sigma}\right)^{-1/\xi} & \text{si} \quad \xi \neq 0 \\ 1-\exp\left(-\frac{x-\mu}{\sigma}\right) & \text{if} \quad \xi = 0 \end{cases}$$

where parameters ξ, μ and σ are defined in the same way as for the GEV function.

For a threshold u, close to the upper bound of the distribution support, function $F(x)$ can be written approximately as:

$$F(x) = \text{Prob}\{X \leq x\} \approx 1 - \text{Prob}\{X > u\} \left(1 + \xi \frac{x - u}{\sigma + \xi\,(u - \mu)}\right)^{-1/\xi}$$

for any value of x such that $1 + \xi \frac{x-u}{\sigma+\xi\,(u-\mu)} > 0$. We then need to estimate parameters based on a function of X, which must contain a number of elements greater than u to produce a reasonable estimation; the value of u must be relatively high. The maximum likelihood criterion is used in this case.

This approach allows us to model distribution tails, which are "heavier" than those found following the application of a Gaussian distribution to the sample. It requires a sufficient number of realizations above the threshold u.

An extension to vectorial cases is possible, and work has also been carried out on modeling the extreme values of a random process, for example [SCH 03]. The notion of copulas, discussed in section 2.1.3.3, allows us to take account of dependencies of extreme values [ACH 12].

2.1.5.2. *Acceptability of an excursion into Extremistan*

The theory of extreme values extends knowledge based on known and accepted realizations below a threshold u and on known, but not necessarily accepted, realizations above u. It is then used to define the probability of events that have never been observed by extrapolation, and the very weak probability level is considered to be acceptable by the decision maker. This is equivalent to defining potential extreme events based on background noise.

The probabilistic approach may be considered as the opposite of the robust design perspective. In section 2.1.2.3, we demonstrated how the border between acceptable and unacceptable events is defined, with the presumption of continuous evolution between the two events. Acceptability consists of defining the size of the convex function and taking all the measures needed to make unacceptable events impossible. This does not constitute an analysis of uncertainty, but an analysis of the global system, something for which a variety of methods are available (for example failure mode, effects and criticality analysis (FMECA)). This provides us with the means of closing the borders of Extremistan.

However, the closure of a border is not certain, and when extreme events may happen, we cannot say that the risk – the combination of the occurrence of an undesirable event and serious consequences – is admissible; a risk is only admissible if the consequences are accepted, whatever the probability of its occurrence may be.

2.1.6. *Conclusion*

This exploration highlights the efforts required from a mechanical perspective in the acquisition and analysis of data.

– Data acquisition: samples are always small and incomplete, but this may be offset by capitalizing on feedback. This is an essential activity for development within companies, which generally have access to a large number of measurements generated by production or maintenance controls. Thus, simple knowledge of the coefficient of variation of a procedure is both a measure of quality and skills, and a particularly relevant value for use in designing new products.

– Data analysis: a simple sample based on a Gaussian distribution is always insufficient, and overhasty modeling leads us to the introduction of arbitrary information. The main information processing tools described in this section have reached a certain level of maturity, and those involved in the sphere of mechanics should make full use of them in the design process, while remembering that a good mathematical solution does not always constitute a good solution from an engineering perspective.

In this section, we have also seen that while uncertainty treatment methods based on mean or median behaviors (Mediocristan) can profit from probabilistic or other approaches, on the condition that sufficient rigor is used in the construction of the stochastic model, extreme events can only be treated in a probabilistic manner on the border between Mediocristan and Extremistan. Extreme events that belong to the territory of Extremistan must be treated using systematic analysis, with the acceptance or refusal of the consequences of failure of each component, whatever the level of probability involved.

2.2. Uncertainty in behavior models

Mechanical design is based on the use of tests and on models. The purpose of these models is to produce a numerical representation of the behavior of a system, defined as an element of the environment in interaction with a user. The system reacts as a function of input data and, in response, produces a scalar or vectorial variable of interest Y. Let $\{X\}$ be the data vector, $\mathbf{M}$ the system state and Y the expected response or observation. We obtain:

$$Y = \mathbf{M}(\{X\}) \qquad [2.1]$$

The designer must then decide whether or not to accept the response, and be able to generate evolutions in the response as required. This is dependent on the representations chosen for both the vector $\{X\}$ and for $\mathbf{M}$, which gives the best possible representation of the physical behavior of the system. Each of these aspects is subject to uncertainty. In this section, we will begin by examining the model of $\{X\}$, placing the description given in section 2.1.2 in the context of the design process; we will then consider the case of $\mathbf{M}$.

2.2.1. *Uncertainty and input data*

[4]The early phases of the design process, particularly those known as "conceptual design" (following the abstraction hierarchy proposed by Rasmussen *et al.* [RAS 94], where data are processed at abstract levels of representation), represent 5% of the design process, but represent more than 75% of the overall cost of the product, according to Carrubba, cited by Sharpe [SHA 95]. Assistance in these phases is, thus, of both ergonomic and economic importance [DAR 01]. Information qualification is one of the possible means of assisting with the management of modifications, filtering the modifications for propagation between different models according to priority. Maturity levels are often used to qualify information used in the design process. These may be defined as the degree of improvement of processes using a predefined set of process domains, in which all of the objectives of the whole project are attained [CHR 07]. Modeling the uncertainty, trust and maturity criteria associated with reused data and/or knowledge presents significant difficulties. Two types of approaches may be taken when modeling uncertainty in the product predesign stage, from the problem definition process to conceptual design.

2.2.1.1. *Qualitative approaches*

Qualitative approaches are based on the concept of preliminary information, introduced by Clark and Fujimoto [CLA 91] to permit parallel

4 From a contribution by S. CASTRIC.

execution of activities in the course of the product development process. Later, Eppinger *et al.* [EPP 97] defined the term "preliminary information" as a parameter that is in constant evolution until a final value is reached. The status of the parameter in its evolution is connected with its maturity [HAN 97].

The qualification and characterization of models and information involve a number of different aspects, including durability, variation, sensitivity and completeness. The design information used by a design office may be classified by durability levels [GRE 07], i.e. the longevity of the information. A scale from "1" to "5" is used in attributing a degree of information validity in relation to a given object. The classification shown in Table 2.1 represents the level of durability and the corresponding qualification.

Durability levels	Qualification
1	Non-durable information
2	Information valid for approximately 1 week, until next modifications
3	Information valid for the duration of the study, approximately 6 months
4	Information valid for several programs
5	Information valid for technologies currently in use

Table 2.1. *Durability levels [GRE 07]*

Table 2.2 shows the different variation quantification levels, i.e. the probability that information will reach its final value. These variation levels were developed by Krishnan [KRI 96], and go from "0" for very unstable to "3" for stable variation, meaning that there is a high probability that the object will approach its final value.

Yassine *et al.* [YAS 99] classified sensitivity levels, i.e. the impact of changes on information, using a scale from "0" to "3", from information considered to be "insensitive" to information considered to be "sensitive". This classification is shown in Table 2.3.

The final classification shown here is the completeness level. This represents the combination of depth (nature of change) with the scale of the information in question. The depth represents the nature of the change acting on the object (imprecision, abstraction and level of detail). The scale is the importance of an item of information in relation to the state of development expected by the user. The completeness represents the evolution of these two dimensions. This scale is presented in Table 2.4, from “0” to “3”. The completeness levels shown in this table were proposed by Yassine *et al.* [YAS 99].

Variation levels	Qualification
0	Very unstable: almost zero probability of an object approaching its final value
1	Unstable: weak probability of an object approaching its final value
2	Moderately unstable: slightly high probability of an object approaching its final value
3	Stable: high probability of an object approaching its final value

Table 2.2. *Activity variation levels*

Sensitivity decisions	Qualification
0	No sensitivity: the activity is insensitive to changes in the input object
1	Weak sensitivity: the activity is not particularly sensitive to changes in the input object
2	Moderate sensitivity: the activity is moderately sensitive to any changes in the input object
3	High sensitivity: the activity is sensitive to any changes in the input object

Table 2.3. *Information sensitivity levels (based on [KRI 96])*

Figure 2.1 [GRE 06] shows the two processes summarizing information maturity. In the first process, the uncertainty of the information is characterized by the emitter. Uncertainty characterization provides responses to a number of questions, concerning the nature, the expected frequency and the level of change. The response to these questions is based on the instability or degree of evolution of the information [KRI 96, YAS 99, TER 99]. The degree of trust placed in the information is symptomatic of the emitter’s degree of knowledge of the information in question.

The second process, information qualification, is the evaluation of the use/validity of the information by the emitter. This process makes use of the relevance, completeness and trust levels described above. These levels are used to respond to questions concerning the origin of the information (has it been produced/transmitted by an expert?), its response to objectives determined by the user and the risks involved in using the information.

Completeness levels	Qualification
0	Incomplete: the object has zero depth or zero scale
1	Very limited: the object has low depth and scale. (The object does not respond to the majority of expectations and the majority of its parts have not been finalized)
2	Partial: the object has moderate depth and scale. (The object responds to the majority of expectations and most of its parts have been finalized)
3	Complete: the object has high depth and scale. (The object responds to all expectations and all parts have been finalized)

Table 2.4. *Information completeness levels [YAS 99]*

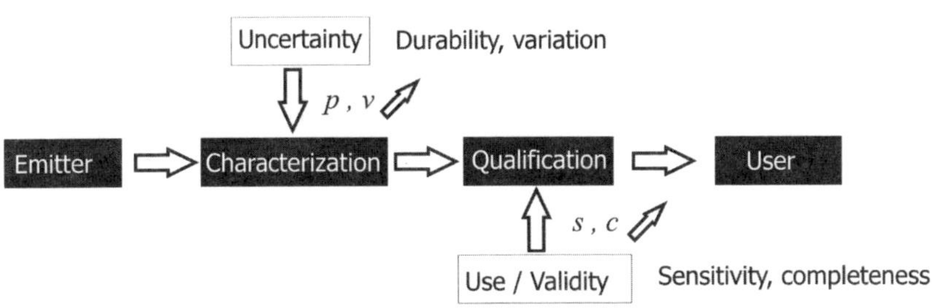

Figure 2.1. *Information uncertainty, from the emitter to the receiver [GRE 06]*

The approaches used in this section are of a qualitative nature. Mathematical tools for quantitative representation were discussed in section 2.1.2 □.

2.2.1.2. *Conclusion*

The knowledge available for an item of data is represented by the qualitative measures described above and by the parameters of the quantitative measures discussed in section 2.1.2. A wide variety of representations are possible, from the simple model based on a scalar, which simply imposes an unjustified precision level and removes the possibility for doubt, to more or less complex models, which attempt to show only genuinely available information. These representations may be grouped into a new object, the *knowledge entity* associated with a variable. Each parameter of the object, and the object as a whole, is subject to an assessment based on scales of durability, sensitivity and variation. Table 2.5 gives a very partial representation of the knowledge entity of a variable X. Each of the models considered – in this case, scalar, possibilistic and probabilistic – is dependent on a certain number of parameters, each associated with a quadruplet (d = durability, v = variation, s = sensitivity and c = completeness) representing the quality of the information for each parameter. The maturity of the variable is itself a function $f(d, v, s, c)$. Clearly, this value increases based on the different proposed models. The construction of tables of this type was suggested by Lemaire [LEM 10] to characterize a knowledge entity, but the design community needs to work together to create a shared table.

Varable	Model	Maturity	Parameters	Maturity
$X^{(k)}$	Scalar	$k = f(d, v, s, c)$	Value	d, v, s, c
$X^{(l)}$	Possibilistic	$l = f(d, v, s, c)$	Possibility $\prod(X)$	d, v, s, c
			Necessity $\mathrm{N}(X)$	d, v, s, c
$X^{(m)}$	Probabilistic	$m = f(d, v, s, c)$	Probability law $F_X(x)$	d, v, s, c
			Mean m_X	d, v, s, c
			Confidence interval	d, v, s, c
			...	p, v, s, c
$X^{(n)}$	$\vdots$	$n = f(d, v, s, c)$	$\vdots$	$\vdots$

Table 2.5. *Defining the body of knowledge of a variable*

Thus, the maturity of a variable develops over the course of the design process, and information matures as a result of the motors shown in Figure 2.1 until it reaches a sufficient level:

$$X^{(k)} \longrightarrow X^{(l)} \longrightarrow X^{(m)} \longrightarrow X^{(n)}, \quad k < l < m < n$$

We now need to generalize this approach to variables, which are functions of space and time. The limit to aim for is an almost certain description in which all of the parameters of $X^{(5,3,3,3)}$ would present maximum values ($d = 5$, $v = 3$, $s = 3$ and $c = 3$).

2.2.2. *Uncertainty in behavior models*

2.2.2.1. *Definition and limits of behavior models*

[5]In formal terms, the behavior model $\mathbf{M}$ connects the response Y (which is scalar in this case, but a vectorial response is also possible) to input data $\{X\}$. It presents the knowledge of physical realities, resulting from reasoning, algorithms or experimentation, in the form of a mathematical model. Relationship [2.1] can be made more precise, noting that the model also depends on parameters $\{P\}$ that are independent of the input data:

$$Y = \mathbf{M}(\{P\}, \{X\})$$

The following definition is proposed:

> *A behavior model is a model deduced from knowledge that, using a law or function, can be generalized, in predictive terms, to all experiences in the established domain, independently of the system in question.*

The first consequence of this definition in terms of model assessment is its capacity for generalization beyond the domain used for identification. It therefore operates by extrapolation, validated by recalibration. This can be illustrated using the gravitational model, based on Newton's law, which is valid whatever the mass in question; it is not necessary to measure the gravitational constant for each use. Its value is one of the parameters $\{P\}$ of a model and, at a certain level of observation, can even be an uncertain data value. Thus, CODATA [COD 13] proposes a value of $G = 6.67384.10^{-11} \pm 0.00080.10^{-11}\,\mathrm{m}^3\,\mathrm{kg}^{-1}\,\mathrm{s}^{-2}$, introducing model-related uncertainty, even in the case of so-called universal parameters. This value is independent of the system we seek to model.

The second consequence is that a physical behavior model will be valid up to the point where it is invalidated by experience. The limit of Newton's theory of gravitation is linked to the speed of light; for experiments involving relativist speeds, Newton's theory no longer applies. A behavior model is thus valid not for a system, but for the domain in which it has been established.

---

5 From a contribution by S. CASTRIC.

With the exception of the standard physical model, which is certainly the most complete model available to us, few behavior models are universally applicable. At best, they remain applicable as long as they have not been invalidated by experience. In a mechanical design context, behavior models cannot be generalized to the same extent and their usage frameworks are often limited. All behavior models must remain compatible with recognized principles, such as the laws of thermodynamics. This situation is the result of technological development, which prevents us from establishing more general physical laws, and the increasing complexity of systems involving multiple technical domains.

2.2.2.2. *Physical modeling*

Using experience, we are able to propose models based on hypotheses, physical tests and experimental feedback. At this level, models have a general utility and apply to phenomena rather than to systems. This is a particularly mature area of research, and one that offers numerous design tools: linear or nonlinear behavior, damage, fatigue and failure mechanics, etc. The full extent of research in this domain will not be discussed in detail here.

The mathematical representation of a model translates hypotheses into operators and parameters that are adjusted to fit with experience, generally solved using numerical algorithms. We therefore encounter three sources of uncertainty:

– Uncertainty concerning hypotheses, requiring careful definition of domains of validity, for example the hypothesis of linearity for small movements.

– Uncertainty concerning the identification of parameters. It is very often possible to find the best-identified parameters to validate a model in relation to an experience, but this becomes harder in the case of multiple experiences.

– Uncertainty concerning the precision of the numerical solution.

Precise hypotheses may be used to form a robust whole leading to a complex mathematical model. This is the case when using a basis of differential or partial differential equations with boundary conditions. A simple example is given by the heat equation connecting temperature θ in domain D with source r and border Γ:

$$\begin{aligned}\Delta\theta &= r \quad \theta \in D \quad \theta_{\Gamma_1} = \theta_1 \text{ and } \left.\frac{\partial\theta}{\partial n}\right|_{\Gamma_2} = \theta_n \\ \theta &= \mathbf{M}(\{X\} = \{r, \theta_1, \theta_n, D, \Gamma\}^t)\end{aligned}$$

A model of this type does not contain parameters and its uncertainty depends solely on the validity of the hypotheses used. These may be insufficient, in which case we must postulate a form of response of the type:

$$\theta = \sum \{\mathbf{F}, \{P\}\}^t \{X\}$$

where $\{\mathbf{F}, \{P\}\}$ constitutes a functional basis on which we project the response (polynomials, neural networks, spline functions, etc.). This model contains parameters that need to be identified experimentally. Care is needed to avoid overlearning, which makes models precise in a limited domain to the detriment of their capacity for generalization. For this reason, it is useful to divide the learning base into two parts: one to be used for identification and the other for validation. It is then possible to obtain estimators for parameter uncertainty using a Bootstrap-type technique (see 2.1.3.5). This method requires a sufficient number of experimental iterations and can be very demanding in terms of calculation resources.

Uncertainty concerning numerical solutions is essentially the result of iterative algorithms and the criteria used to validate their convergence. Solutions are produced by approximations, as in the case of finite elements, by framing energy. These approximations contribute to demonstrating the convergent character of results: see [LAD 08], for example. The question of the numerical precision of 64 bit coding, i.e. to approximately 14 significant figures, does not appear to be critical; this level of precision is considerably greater than that of the data, and if a numerical robustness problem was to appear, it would demonstrate the presence of a non-robust design.

2.2.2.3. *Phenomenological modeling*

When physics alone does not provide all of the information required, we must use a complex model containing both design variables and parameters. These modelings involve models of phenomena from a wide range of disciplines and approaches: mechanics, chemistry, thermodynamics; solids, liquids, gases, etc. For illustrative purposes, and without going into detail,

Hardenberg *et al.*, cited by Heywood [HEY 88], carried out tests on a direct injection diesel motor. They proposed the formulation of an output variable based on both a behavior model of a phenomenon – the Arrhenius law – and parameters:

$$\begin{aligned} Y &= (0.36 + 0.22\,Sp).e^{\left[E_a\left(\left(\frac{1}{R\,T}\right)-\frac{1}{17190}\right)\left(\frac{21.2}{(P-42.4)^{0.63}}\right)\right]} \\ \{P\}^t &= \{0.36,\ 0.22,\ 17190,\ 21.2,\ 42.4,\ 0.63,\ R\} \\ \{X\}^t &= \{Sp,\ Ea,\ T,\ P\} \end{aligned}$$

This formulation includes the design variables $\{X\}$ and the "constant" numerical parameters resulting from identification. The parameter R is the ideal gas constant.

The domain of validity of the model remains limited. Heywood defined his approach for the case of diesel monoinjection and for motors with injection pressures much lower than those in current use. The model does not, therefore, appear to be suited to use with current technologies. However, the form of the model is still used as a generally accepted base for study by motor engineers [BOU 10]. This is due to the context in which the models were developed. From an industrial perspective, it is difficult, if not impossible, to wait for a behavior model to be validated and generalized for all innovations. Moreover, the reliance on multiple technical domains means that not all of the required measurements are accessible, such as the temperature or movements of air in the chamber. The behavior model defined by Heywood is still the closest available model to reality, but requires recalibration.

This example illustrates the complexity involved in constructing behavior models. Clearly, once these models have been established in a way suitable for generalization, they offer considerable benefits in the design cycle. The associated uncertainty is mastered as the constants used have been identified and may be modified in line with technological advances.

However, the design process involves a range of technical domains and is subject to considerable evolution over time. Behavior models do provide knowledge, but their accuracy is relative. Systems evolve too quickly, and the necessary measurements are not always available. Furthermore, in this case, we see that the variability of these constants is not taken into account. The quality of tests and uncertainty linked to measurements are often the points with the greatest impact in the case of behavior models. The quality of these aspects determines the generalization capacity of the model. In the following section, we will consider the management of data and model maturity.

2.2.2.4. *Improvement and recalibration of models*

The identification of model parameters is validated for a given domain; input data within, or even from outside, this domain have an impact on identification.

Often, the general law is not called into question, but constants are re-identified in the course of the design cycle. This is known as recalibration. We might, therefore, wonder why physical models, which do not require this recalibration, have not been defined.

The principle of "updating" models is related to a lack of knowledge and to uncertainty concerning the hypotheses of "physical" theories, in the broadest possible sense. For calibration-free models, these hypotheses would therefore need to be extended, adding complexity to the model by including new phenomena, or moving from a two- to a three-dimensional (3D) model. However, even if all of the available knowledge resulting from experience were to be taken into account, it would still be impossible to generalize the model due to the lack of test-based information; furthermore, tests are not always sufficiently representative of real operating conditions. A behavior model is identified using a system resulting from a technology, and is therefore specific. To obtain a global model, we would need to compare several data sets from different systems and different technologies. From an industrial perspective, this approach is not realistic for competitive and economic reasons.

Another reason for recalibration is model complexity. 3D models, while we may presume that they allow a higher level of generalization and thus reduce the need for recalibration, also require significant computation time and this is not always compatible with the time constraints of design. In such cases, it is better to use a simpler, recalibrated modeling, switching from the 3D model to a simpler alternative.

Recalibration necessitates additional tests for a model to be adapted to the evolutions of a system during the design process. The interest of a physical approach lies in the fact that the need for testing during the learning phase is limited by capitalizing on prior knowledge of the system. A Bayesian approach is often used: in this case, it is possible to integrate different types of knowledge, whether qualitative or quantitative [BER 92], and the parameter vector $\{P\}$ thus becomes random.

The border between behavior models and black box models also requires attention. In this context, we encounter the notion of gray box, or hybrid, modeling. In this type of approach, the level of knowledge allows us to divide a system into subsystems; for each of these subsystems, a modeling approach is defined by what we know. The objective is to limit the number of tests while guaranteeing maximum predictive capacity. As we have seen, in certain situations, the theory is not sufficiently advanced to allow the development of a behavior model. In this case, the division of the problem allows us to use hybrid approaches, mixing statistical and knowledge-based models.

The main difficulty involved in using a hybrid approach resides in obtaining intermediary information. It is often impossible to measure the internal variables of a system: once again, the problem is not created by the model itself, but by the knowledge and precision of input data. What level of uncertainty should we attach to a thermodynamic temperature model based on the ideal gas law? Is the error – or deviation – acceptable in this case? Is this model more precise than an identification using a purely mathematical approach? The answers to these questions depend mainly on the study environment and the conditions of use of the model. The major benefit of the physical model is its strong basis in reality. It is then possible to compare the results obtained using this model to those for a similar system, or to compare two different behavior models □.

2.2.2.5. *Observation models*

The models discussed in the two previous subsections are based on physical knowledge and recalibration. From this perspective, uncertainty regarding $\{X\}$ is introduced by the deterministic parameters $\{P\}$ of a physical modeling. In the absence of a physical hypothesis to govern the construction of a behavior model [2.1], we may represent knowledge directly using a subjacent mathematical structure, enriched by observation. Two approaches are possible: using Markov processes or Bayesian networks. These approaches are both based on the description of relationships between variables. From this perspective, they allow us to represent systems that are much more complex than the simple relationship [2.1].

2.2.2.5.1. Markov processes

[6]A first-order, continuous or discrete time Markov process [BER 98, BRE 09] is a sequence of random variables, $\{X\} = \{X_{t_1}, X_{t_2}, ..., X_{t_n}\}$, for any finite subset of instants $\{t_i, i = 1, 2, ..., n\}$ such that $t_1 < t_2 <$

6 Reread by C. MATTRAND.

... $< t_n$ takes values in a discrete or continuous space such that the conditional distribution of $X_{t_{k+1}}$, given $\{X_{t_1}, ..., X_{t_k}\}$, is equal to the distribution $X_{t_{k+1}} \,|\, X_{t_k}$. If the indexing relates to time, this implies that the prediction of the future $X_{t_{k+1}}$ depends solely on the knowledge of the present X_{t_k}. The state space may be discrete (Markov chain) or continuous, and describe time or any sequence representing a physical evolution. The first-order model may be extended by the introduction of r variables preceding the present variable. A bivariate process may also be used, with the inclusion of a hidden level [CAP 05].

The distribution of a chain (discretized time of dimension n) is defined by the distribution X_{t_1}, and by the data for the transition matrix $[P]$ of which the elements are the probabilities $p_{i,j} = \text{Prob}\{X_{t_{k+1}} = e_j \,|\, X_{t_k} = e_i\}$ with the consistency condition $\sum_{j=1}^{N} p_{i,j} = 1$, $0 \leq p_{i,j} \leq 1$, for $i = 1, ..., n$. Expert assessment, or, better, a statistical treatment of observed data, is used to complete matrix $[P]$. The quality measures of the statistical estimation (section 2.1.3) are used to validate – or invalidate – the estimated probabilities (variation coefficient, Bootstrap); Berchtold [BER 98] also proposes a stability measure.

An illustration is provided in Mattrand's doctoral thesis [MAT 11, MAT 11a, MAT 11b], which concerns the probabilistic approach of tolerance to fatigue damage. Two types of processes are used to model sequences of fatigue cycles: a first-order chain and a bivariate hidden process model. The variable $X_k = \{M_k, m_k\}$ is a fatigue cycle, where M_k and m_k designate, respectively, the maximum and minimum points of the cycle. The inference is applied to measured cycle data obtained from aircraft in the course of operations. A total of 27,458 flights were grouped into 27 classes representing missions. In this way, a representation of the behavior model $Y = \mathbf{M}(\{X\})$ was obtained without needing to specify the damage source variables $\{X\}$ or the physics $\mathbf{M}$, and validated, at least for classes containing a sufficient number of elements $\square$.

2.2.2.5.2. Bayesian networks [BOR 10]

Bayesian networks are used to represent knowledge of a technical or other system. Existing knowledge is used to construct an oriented graph showing the relationships between random variables, for example $\{X\} \longrightarrow Y$, but the network may be considerably more complex. The relationships between variables are probabilities inferred from observations. The network is, in fact, a *concise representation of the joint probability distribution of variables, which is the most complete information we can provide in connection with this set of random variables* [BOU 13]. These networks serve the same objective as copula theory (see section 2.1.3.3).

The first stage in establishing a Bayesian network is to construct the graph, which provides the structure for the knowledge model. The second stage is inference, i.e. the calculation of conditional probability forecasts based on observed probabilities.

The graph of the behavior model $Y = \mathbf{M}(\{X\})$ shown in Figure 2.2(a) shows that model $\mathbf{M}$ is unknown and connects the random parameters $\{X\}$ with observation Y. We must then use these observations to construct the joint distribution. If the network is limited to this topology, it presents limited interest, and inferences will be limited to Bayesian updating (section 2.1.3.4). A more general form is shown in Figure 2.2(b) [COW 98], cited by [BOR 10], using parameter, latent and observable variables.

In this case, the property of *conditional independence* is essential. A variable Z is conditionally independent of Y in relation to (or given) the information $X = x$ if an information input into Y does not modify the uncertainty relating to Z. Thus:

$$[Z \mid Y = y, X = x] = [Z \mid X = x]$$

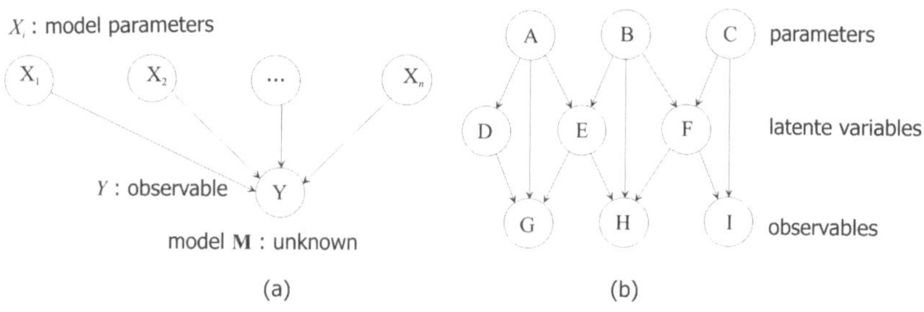

Figure 2.2. *Graphical model of a Bayesian network*

This must be verified, for example, for $\{A = X, D = Y, E = Z\}$ (Figure 2.2(b)). This property shows the Markovian character of the network:

– A node is conditionally independent of its non-descendants given its parents.

– The probability distribution of a node, given the rest of the network, depends only on its parents, its descendants and the co-parents of its children.

A Bayesian network is therefore represented by a directed acyclic graph (DAG). It represents expert knowledge of the relationships between variables, without explicitly illustrating knowledge of the physics of this relationship.

The variables of the graph constitute a random variable vector; by inferring this vector, we aim to identify the joint distribution. Conditional independence allows us to express the joint distribution as a product of marginals (in the case of Figure 2.2(b)):

$$\begin{aligned}\{X\} &= \{A, B, C, D, E, F, G, I\} \\ \text{Prob}(\{X\}) &= \textstyle\prod_{\{X\}} \Pr\left(X \mid \{\text{parent}(X)\}\right) \\ &= \Pr(A)\ \Pr(B)\ \Pr(C) \\ &\times \Pr(D \mid A)\ \Pr(E \mid A, B)\ \Pr(F \mid B, C) \\ &\times \Pr(G \mid A, D, E)\ \Pr(H \mid B, E, F)\ \Pr(I \mid C, F)\end{aligned}$$

Given the information available *a priori*, the inference process consists of calculating *a posteriori* probabilities given the new input information. The problem is NP-difficult (meaning that it is possible to verify a solution, but very difficult to find one), and in addition to exact methods (for simple cases), a range of simulation-based approximation methods may be used (Gibbs sampling, the Monte-Carlo Markov Chain (MCMC), the Metropolis–Hastings algorithm [BOR 10], etc.).

2.2.2.6. *Conclusion*

The construction of a behavior model requires us to establish hypotheses and identify parameters. The uncertainty associated with a model thus depends partly on the trust in the hypotheses, used within a domain of validity and variability of parameters, and on the numerical processes used (64 bit coding of variables and precision of the convergence of algorithms).

Scientific knowledge reduces epistemic uncertainty, and models may be classified based on the breadth of the hypotheses they use. Parameter identification creates aleatoric uncertainty, which, indirectly, includes the bias present in hypotheses. To our knowledge, a model trust scale has yet to be established. The scales defined in section 2.2.1.1 apply to parameters, but only a qualitative hierarchy may be offered in the case of hypotheses.

Wherever possible, we should look for models that allow for a bounding of the desired variables of interest, and not simply a numerical approximation based on a given model. This is something of a vain hope, except in certain specific cases. One example may be found in static and cinematic elastoplastic theories, which operate by framing the value of interest. The question of quantifying the gap (and not the error) of the model in relation to the chosen hypotheses is also still open.

A behavior model $\mathbf{M}$ is defined by a basis function $\mathbf{F}$ with uncertain parameters $\{P\}$ applied to data $\{X\}$. In section 2.3, we will consider the propagation of uncertainty in this model.

2.3. Uncertainty propagation

2.3.1. *The problem of uncertainty propagation*

In the previous sections of this chapter, we have considered ways of processing data to obtain information reflecting true objective knowledge (section 2.1), then the ways of defining models of input variables and of physical behaviors (section 2.2). Input variables $\{X\}$ are represented as part of a *knowledge entity*, as illustrated in Table 2.5. Behavior models are based on physical knowledge and on feedback, expressed using a function $\mathbf{F}$ including parameters $\{P\}$ that are, themselves, uncertain. To summarize:

– $\{X\}$ is the vector of the input variables, the elements of which are deterministic, possibilistic or probabilistic values, intervals or other entities.

– $\mathbf{F}$ is a function reflecting behavior. It is represented either by an algorithm including detailed modeling, by a meta-model resulting from this algorithm or by an approximation using a given basis: polynomials, neural networks, splines, etc. It should be associated with a measurement of the deviation or bias of the model.

– $\{P\}$ is the vector of the parameters used in the behavior model, and is itself represented by a knowledge entity, recalibrated where necessary.

Taking $\{Y\}$ as the vector of quantities of interest produced by the models used in decision-making (constraint levels, dimensioning tolerance, cost, mass, etc.), these quantities of interest can be identified by solving:

$$\{Y\} = \mathbf{M}\left(\underbrace{(\mathbf{F},\{P\})}_{\text{behavior model}},\ \underbrace{\{X\}}_{\text{data model}}\right) \qquad [2.2]$$

By calculating $\{Y\}$, we construct the associated knowledge entity as a function of the knowledge entities of $\{P\}$ and $\{X\}$ for a given $\mathbf{F}$. In general terms, we require an algebra of the knowledge entities; this algebra does not exist. However, a considerable and varied body of published work exists for cases with a homogeneous entity.

In the extreme case, uncertainty is absent, and $\{X\}$ and $\{P\}$ only contain deterministic (best estimate) values; in this case, the calculation of deterministic values of $\{Y\}$ is conceptually simple, although it may still be highly demanding in terms of resources. This point is important as all calculations of uncertainty involve judicious repetitions of deterministic calculations. We will consider two classes of questions:

– Given the uncertainties of $\{X\}$ and $\{P\}$, what uncertainties are involved in $\{Y\}$? This question relates to the domain of *sensitivity analysis* and is dealt with by constructing entity $\{Y\}$.

– Given that there is an admissible domain for $\{Y\}$, how do we construct the separator between acceptable and unacceptable realizations? In this case, we add an acceptability criterion to the uncertainty analysis; the question relates to the domain of *reliability analysis*.

In this section, we will first consider a number of homogeneous transmission models used in sensitivity analysis, before looking at the question of classification in reliability analysis and the use of model reduction in limiting the required calculating power. We will then offer a general overview of uncertainty quantification (UQ).

2.3.2. *Analyzing sensitivity to uncertainty*

2.3.2.1. *Simulation methods*

A realization of a component Y of the output variable can be calculated as soon as we have realizations of $\{X\}$ and $\{P\}$. We must therefore simply repeat the calculation for all possible realizations while respecting the nature of the knowledge entity. Thus, in the case of interval variables, we must cover the set of intervals, then identify the bounds of the output variable. If the knowledge entity is probabilistic, the simulation generates a sample according to the distribution of the variable. This is the classic Monte-Carlo method [RUB 81].

Depending on the form of the behavior model, simple relationships may be used to avoid processing a large number of simulations, based on interval, possibility or probability algebra. An example of this type may be seen in the linear combination of Gaussian variables. If X_1 and X_2 are Gaussian, then

$a_1 X_1 + a_2 X_2$ is Gaussian. However, there is no reason why physics, even less mechanics, should be linear and Gaussian.

2.3.2.2. *Interval models*

2.3.2.2.1. Interval arithmetic

[7]An interval $\mathbf{x} = [\underline{x}, \bar{x}]$, $\underline{x}, \bar{x} \in \mathbb{R}$ is a set of real numbers x such that $\underline{x} \leq x \leq \bar{x}$. Interval arithmetic was introduced by Moore [MOO 79]. The elementary operations are defined in the following manner:

$$\mathbf{x} \oplus \mathbf{y} = [\underline{x} + \underline{y}, \bar{x} + \bar{y}]$$

$$\mathbf{x} \ominus \mathbf{y} = [\underline{x} - \bar{y}, \bar{x} - \underline{y}]$$

$$\mathbf{x} \otimes \mathbf{y} = [\min(\underline{x}\,\underline{y}, \underline{x}\,\bar{y}, \bar{x}\,\underline{y}, \bar{x}\,\bar{y}), \max(\underline{x}\,\underline{y}, \underline{x}\,\bar{y}, \bar{x}\,\underline{y}, \bar{x}\,\bar{y})]$$

$$\mathbf{x} \oslash \mathbf{y} = [\underline{x}, \bar{x}] \otimes [\frac{1}{\bar{y}}, \frac{1}{\underline{y}}] \quad \text{if} \quad 0 \notin [\underline{y}, \bar{y}]$$

where $\oplus$, $\ominus$, $\otimes$ and $\oslash$, respectively, represent the operations of addition, subtraction, multiplication and division between the interval variables $\mathbf{x}$ and $\mathbf{y}$.

It is important to note that these operations do not respect classic properties: for example, $[0,1] \otimes (1-1) = [0,0] = 0$ while $[0,1] \ominus [[0,1] = [-1,1]$. Multiplication is not distributive, which can lead to overestimation in calculations. However, subdistributivity is maintained. It is also important to note that $\mathbf{x} - \mathbf{x} \neq 0$ and $\frac{\mathbf{x}}{\mathbf{x}} \neq 1$. Moreover, when a given variable occurs more than once in an operation, it is treated as a different variable for each occurrence, which can lead to overestimation due to dependency problems. Interval arithmetic thus provides us with intervals that are certain, but often overly conservative [KRE 96].

2.3.2.2.2. Affine arithmetic

Affine arithmetic was introduced in 1993 by Comba and Stolfi [COM 93] and developed further by De Figueiredo and Stolfi [FIG 04]. This method allows us to retain traces of correlations between quantities during calculations. This additional information provides much stricter enclosures in the calculation process, avoiding the compounding of errors. Each quantity x is represented by an affine form $\hat{x}$, which is a first-degree polynomial function of noise variables:

$$\hat{x} = x_0 + x_1 \varepsilon_1 + ... + x_n \varepsilon_n$$

7 From a contribution by F. AGUIRRE.

The noise ε_i is a real symbolic variable with a value in the interval $U = [-1, +1]$ and is independent of the noises $\varepsilon_{j \neq i}, \forall j$. The coefficient x_0 is taken as the central value of $\hat{x}$. The coefficients $x_1, ..., x_n$ are known as partial deviations, associated with the noises $\varepsilon_1, ..., \varepsilon_n$ in x. Conversions between affine forms and intervals are carried out in the following manner:

– Affine $\rightarrow$ Interval:

$$\hat{x} = x_0 + \sum_{i=1}^{n} x_i \, \varepsilon_i \longrightarrow \mathbf{x} = \left[x_0 - \sum_{i=1}^{n} |x_i|, \, x_0 + \sum_{i=1}^{n} |x_i| \right]$$

– Interval $\rightarrow$ Affine:

$$\mathbf{x} = [\underline{x}, \, \bar{x}] \longrightarrow \hat{x} = \frac{\underline{x} + \bar{x}}{2} + \frac{\bar{x} - \underline{x}}{2} \varepsilon_1$$

The extension of non-affine interval operations requires a good approximation of the exact result. An additional term is also needed to reflect the error inherent in this approximation. Different approximation methods may be used for the affine forms of several functions: Chebyshev, min-range and interval approximation [ROC 04, LEE 06].

2.3.2.2.3. Set-based approach to robust design

[8]Interval-based modeling allows for completely robust design if the bounds used are certain. Set-based approaches identify the domains of design variables that satisfy a set of performance objectives despite model uncertainties, uncertainty relating to design variables and noise factors.

To formalize this approach, Qureshi [QUR 11] provides fundamental definitions of a model based on formal logic. $\mathcal{V}$ represents the set of design variables, $\mathcal{U}$ represents the set of uncertainties, $\mathcal{D}_{\mathcal{V}}$ represents the initial search domain for $\mathcal{V}$, $\mathcal{D}_{\mathcal{U}}$ represents the domain of uncertainties and $\mathcal{C}$ represents the set of model constraints. These may be continuous, discrete, complex or relational. If a model contains n constraints, the set $\mathcal{C}$ will be written $\mathcal{C} = \{c_1, ..., c_n\}$. Certain constraints may result from meta-models constructed using experimental data or numerical simulations; in this case, uncertainties are estimated and taken into account in vector $\mathcal{U}$.

---

8 From a contribution by J.-Y. DANTAN.

This type of formalization requires the use of logical quantifiers, such as $\forall$, $\exists$, etc., applied to the elements of $\mathcal{V} \rightarrow \mathcal{Q}_{\mathcal{V}}$ and $\mathcal{U} \rightarrow \mathcal{Q}_{\mathcal{U}}$ to express the conditions of acceptability of robust solutions in the set $\{\mathcal{Q}_{\mathcal{V}}, \mathcal{Q}_{\mathcal{U}}, \mathcal{D}_{\mathcal{V}}, \mathcal{D}_{\mathcal{U}}, \mathcal{C}\}$.

A solution is considered to be robust if it belongs to the set of solutions where the constraints are satisfied, for all affectations of design variables and for all affectations of uncertainties, within their own domain.

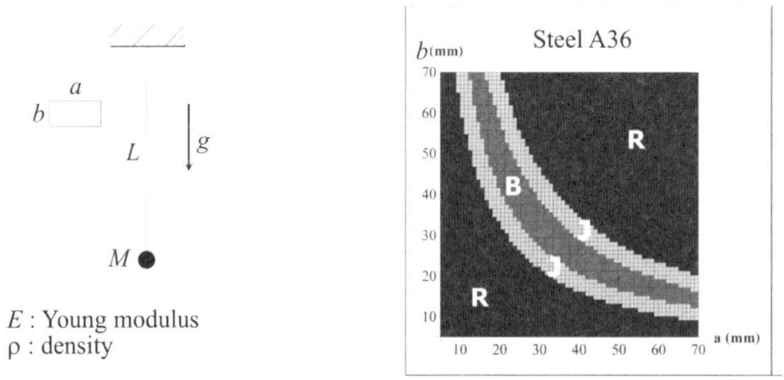

Figure 2.3. *Illustration of a robust design approach using interval arithmetic*

2.3.2.2.4. Illustration

Qureshi gives a simple illustration, Figure 2.3, briefly described below, alongside technical applications. A bar of length L, with a rectangular cross-section with sides a and b, placed in a gravitational field g, supports its own weight and mass M. The following conditions are required:

– the normal stress, the mass of the bar and its lengthening ΔL must be lower than fixed admissible values;

– dimensions $\mathbf{a} = [\underline{a}, \bar{a}]$ and $\mathbf{b} = [\underline{b}, \bar{b}]$ are interval variables;

– the other data values are scalar.

Tolerancing involves uncertainty relating to dimensions $\mathbf{a} \in [\underline{a}, \bar{a}]$ and $\mathbf{b} \in [\underline{b}, \bar{b}]$ and to L. These are noted using the intervals $\delta\mathbf{a}$, $\delta\mathbf{b}$ and $\delta\mathbf{L}$. Using his chosen numerical data, the author obtained the result shown in Figure 2.3 (right), showing three defined domains. Domain **R** contains inadmissible solutions. It is made up of the bi-dimensional intervals $\mathbf{a} \times \mathbf{b}$, which do not satisfy the prescribed constraints. Domain **B** contains admissible solutions, and domain **J** contains intervals for which no conclusion can be reached. Domain **B** illustrates the notion of a robust interval, as defined in section 1.2.3 □.

2.3.2.3. *Possibilistic models*

2.3.2.3.1. Fuzzy numbers

Possibilistic models make use of fuzzy number measures in the knowledge entity. We therefore need to consider elementary operations involving these numbers. A fuzzy number $\tilde{n}$ is a fuzzy subset of the real numbers of which the numbers with the highest degree of membership are grouped around a real number, known as the mean value [DUB 78]. The membership function has the following characteristics, as illustrated in Figure 2.4:

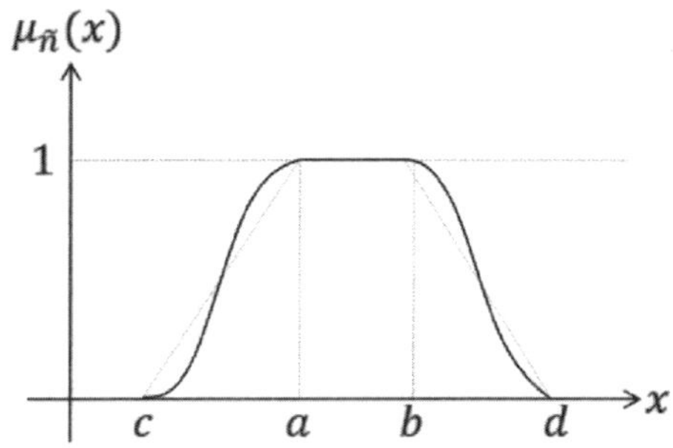

Figure 2.4. *Representation of a fuzzy number*

– monotony on each side of the mean value;

– continuous representation of $\mathbb{R}$ toward the interval $[0, 1]$;

– constant in the interval: $]-\infty, c] : \mu_{\tilde{n}}(x) = 0 \; \forall x \in]-\infty, c]$;

– increasing in the interval: $[c, a]$;

– constant in the interval $[a, b]$: $\mu_{\tilde{n}}(x) = 1 \; \forall x \in [a, b]$;

– decreasing in the interval: $[b, d]$;

– constant in the interval: $[d, +\infty[\; : \mu_{\tilde{n}}(x) = 0 \; \forall x \in [d, +\infty[$,

where a, b, c and d are real numbers. The following specific cases may also occur:

$$c = -\infty, \text{ or } a = b, \text{ or } c = a, \text{ or } b = d, \text{ or } d = +\infty$$

The mean value m of the fuzzy number is an element of $[a, b]$, generally $m = (a + b)/2$. Certain specific cases should be taken into consideration.

– If $a = b = c = d$, the fuzzy number represents an ordinary real number.

– If $a = b$, the membership degree of the fuzzy number is "approximately a".

– $\mu_{\tilde{n}}$ is the truth value of the statement "the value of $\tilde{n}$ is x".

– Two fuzzy numbers are equal if, and only if, they have the same membership function.

2.3.2.3.2. LR (trapezoidal) fuzzy numbers

An "LR" description may be used to facilitate operations [DUB 78]. This involves dividing a trapezoidal fuzzy number into three sections: the left (L), the right (R) and the mean.

L is the restriction of the fuzzy number $\tilde{n}$ in the interval $[c, a]$, and is continuous and strictly increasing within its interval. R is the restriction of the fuzzy number $\tilde{n}$ in the interval $[b, d]$, and is continuous and strictly decreasing in its interval. Thus:

$$\mu_{\tilde{n}} = \begin{cases} L\left(\frac{a-x}{\beta}\right) & c < x < a,\ \beta > 0 \\ 1 & a \leq x \leq b \\ R\left(\frac{x-b}{\delta}\right) & b < x < d,\ \delta > 0 \end{cases}$$

where β and δ are normalization constants.

With fuzzy number of this type, elementary operations such as addition and multiplication do not depend on the analytical expressions of L and R, but on a, b, c and d alone if the same type of function is used for all of the fuzzy numbers and the output value has the same membership function.

2.3.2.3.3. Elementary operations using trapezoidal fuzzy numbers

Let us take two fuzzy numbers $\tilde{m}$ and $\tilde{n}$ with $a = b$, defined by:

$$\tilde{m} = (m,\ \alpha,\ \beta) \quad \text{with :} \quad \mu_{\tilde{m}} = \begin{cases} L\left(\frac{m-x}{\alpha}\right) & c < x < a,\ \alpha > 0 \\ 1 & a \leq x \leq b \\ R\left(\frac{x-m}{\beta}\right) & b < x < d,\ \beta > 0 \end{cases}$$

$$\tilde{n} = (n,\ \gamma,\ \delta) \quad \text{with :} \quad \mu_{\tilde{n}} = \begin{cases} L\left(\frac{n-x}{\gamma}\right) & c < x < a,\ \gamma > 0 \\ 1 & a \leq x \leq b \\ R\left(\frac{x-n}{\delta}\right) & b < x < d,\ \delta > 0 \end{cases}$$

The elementary operations between these two numbers are defined in the following manner:

– Fuzzy addition $\oplus$

$$\tilde{m} \oplus \tilde{n} = (m, \alpha, \beta) \oplus (n, \gamma, \delta) = (m + n, \alpha + \gamma, \beta + \delta).$$

– Fuzzy subtraction $\ominus$

$$\tilde{m} \ominus \tilde{n} = \tilde{m} \oplus \widetilde{-n} \qquad \text{with:} \qquad \widetilde{-n} = (-n, \delta, \gamma).$$

$$\tilde{m} \oplus \widetilde{-n} = (m, \alpha, \beta) \oplus (-n, \delta, \gamma) = (m - n, \alpha + \delta, \beta + \gamma).$$

– Fuzzy multiplication $\otimes$

There is no exact solution for multiplying two fuzzy numbers, but an approximate solution may be obtained using:

$$\tilde{m} \otimes \tilde{n} = (m, \alpha, \beta) \otimes (n, \gamma, \delta) \approx (m\, n, m\, \gamma + n\, \alpha, m\, \delta + n\, \beta).$$

– Fuzzy division $\oslash$

There is no exact solution for division using two fuzzy numbers, but an approximate solution may be obtained using:

$$\tilde{m} \oslash \tilde{n} = \tilde{m} \otimes \widetilde{1/n} \qquad \text{with:} \qquad \widetilde{1/n} \approx \left(\tfrac{1}{n}, \tfrac{\delta}{n^2}, \tfrac{\gamma}{n^2}\right).$$

$$\tilde{m} \otimes \widetilde{1/n} \approx \left(\frac{m}{n}, \frac{\delta\, m + \alpha\, n}{n^2}, \frac{\gamma\, m + \beta\, n}{n^2}\right).$$

– Generalization

Trapezoidal fuzzy numbers present certain advantages due to their simplicity, but with the drawbacks associated with approximation. Kaufmann and Gupta [KAU 91] propose a generalization, using the notion of α-cuts. Each α-cut represents the interval of values with a membership degree greater than or equal to α. A fuzzy number is broken down using the set of its cuts, transforming fuzzy arithmetic into interval arithmetic applied to different membership degrees.

2.3.2.3.4. Propagation of possibilistic information

Moving beyond the question of simple arithmetic rules, the general problem of uncertainty propagation in relationship [2.1] remains open for discussion. Baudrit *et al.* [BAU 06, BAU 07] examine the propagation of mixed probabilistic and possibilistic information. The parameters are aleatoric or deterministic, but the information concerning these parameters is incomplete and is represented by possibility distributions. The output variable may be represented by a fuzzy probability or by a probability interval □.

2.3.2.4. *Probabilistic model – polynomial chaos*

[9]Using a probabilistic model, we aim to identify the stochastic response of a function with random variables. With the exception of linear combinations of Guassian variables or products–quotients of log-normal variables with the well-known solutions, Monte-Carlo simulations are used to obtain a response [RUB 81]. This method consists of carrying out a significant number of replications, determined using the probability laws of uncertain parameters, and calculating the deterministic response of the model for each set of parameters. The sample of different values provided by the response is used to determine the statistical moments, or even an output distribution for the model. However, this approach quickly becomes prohibitively expensive in terms of processing power due to the high number of replications required to obtain a satisfactory confidence level, given that the convergence is a function of $\sqrt{n}$, where n is the size of the sample.

Polynomial chaos is a mathematical tool that offers an efficient response when considering uncertainty propagation, and is less expensive than the Monte-Carlo method. The formal principles of polynomial chaos are presented in section 2.1.4. As we have already seen, a random variable or field can be projected onto a basis function $\Psi_j(\xi)$ with random variables $\xi(\omega)$ (for example Gaussian variables $\mathcal{N}(0,1)$ for Hermite chaos):

$$X(x,\xi) = \sum_{j=0}^{\infty} \phi_j(x)\,\Psi_j(\xi) \approx \sum_{j=0}^{p} \phi_j(x)\,\Psi_j(\xi) \qquad [2.3]$$

where p is the truncation order, chosen based on the level of precision required.

The problem is then reduced to the identification of modal coefficients or functions. This can be achieved using two types of methods: intrusive or non-intrusive.

Intrusive methods use Galerkin projection techniques to express the model in the polynomial chaos basis as a system of coupled, deterministic equations, retaining the linear or nonlinear properties of the original model [GHA 91, BAB 04]. Stochastic coefficients are then calculated by solving the deterministic equation system, allowing significant savings in terms of calculation time compared to the Monte-Carlo approach [XIU 03]. However, the in-depth modification of the model entailed by the projection process can

9 From a contribution by S. BERGER.

pose problems, particularly from an industrial perspective, as it requires a bulky coded implementation. When the number of variables, and thus the number of stochastic modes to calculate, is high, the number of equations becomes significant and implementation becomes laborious.

The non-intrusive methods introduced in section 2.1.4 present the advantage of only requiring the response value of the model for a limited number of sets of input parameters, and do not necessitate modification or manipulation of the model and its numerical implementation.

Once the stochastic modes have been calculated, the polynomial chaos expansion gives us an estimation of the response of the model, allowing us to determine statistical moments up to a chosen order and to reconstruct the probability distribution.

As we saw in section 2.1.4, one of the main advantages of using polynomial chaos is the fact that we do not require *a priori* knowledge of the distribution of the response of the model. Its limitations include the inevitable truncation of chaos, which limits us to an approximation of the model response, and the problem of validating the expansion □.

2.3.2.4.1. Probabilistic models and finite elements

The numerical model of the finite elements method is based on a discretization of data fields and variables of interest, leading us to work with vectors: nodal displacements, mechanical properties, etc.

Let us consider a random field in Figure 2.5, for example the elasticity modulus, denoted by $H(x, \omega)$, where x is the space variable and ω is the aleatoric variable. An initial approach would consist of creating a spatial discretization $H(x, \omega) \rightarrow \{H(\omega)\}; [\rho]$, where $\{H(\omega)\}$ is the vector of the values of the field at the discretization points and $[\rho]$ is the correlation matrix. The discretized variables are then directly accessible by any code using finite elements, which may be used with no modifications.

Another technique is to use spectral expansion, separating the random field into an infinite sum of products of a random variable $\xi(\omega)$ by a geometric mode $\phi(x)$ multiplied by a coefficient λ. This expansion is illustrated in Figure 2.5 for $E(x, \omega)$ using the classic Karhunen–Loève expansion [KAR 46]. A vector of the coefficients, $\{\lambda\}$, is therefore used in the place of the vector of physical variables $\{H(\omega)\}$. These parameters are not accessible in codes using finite elements, and their implementation requires the reconstruction of stiffness matrices, inserting geometric modes, and is thus intrusive in terms of the code.

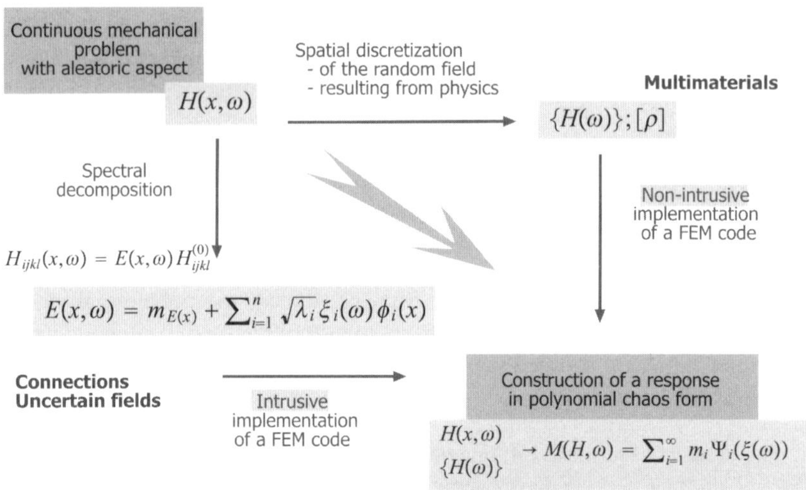

Figure 2.5. *Comparison of intrusive and non-intrusive approaches*

The intrusive method is most efficient as it takes account of the hierarchy of variability modes, but it requires the development of tailored codes. In these cases, we speak of *spectral stochastic finite elements*. The non-intrusive method is easy to implement, and benefits from a firm basis in current coding knowledge. In this case, we speak of stochastic finite elements, and any variable that is accessible in code data can be probabilized. The two pathways should converge as discretizations are refined.

2.3.2.4.2. Multidimensional chaos

Introducing equation [2.3] into relationship [2.2], we obtain the following output variable:

$$\begin{aligned} Y(\{\xi\}) &= \mathbf{M}\left((\mathbf{F}, \{P(\{\xi\})\}), \{X(\{\xi\})\}\right) \\ &= \sum_{j=0}^{\infty} \tilde{a}_{ij}\, \Psi_j(\{\xi\}) \approx \sum_{j=0}^{J-1} \tilde{a}_{ij}\, \Psi_j(\{\xi\}) \end{aligned} \quad [2.4]$$

where:

– $\{\xi\} = \{\xi_1, \xi_2, ..., \xi_m\}$ is a vector of m random variables ($m = \dim\{X\} + \dim\{P\}$), for example centered standardized Gaussians for Hermite chaos;

– each component of $\{X\}$ (respectively, $\{P\}$) is associated with a development of relationship [2.3] using a one-dimensional basis $\phi_j(\xi)$ with coefficients a_j;

– vector $\{X\}$ (respectively, $\{P\}$) is expressed on a basis of m expansions that are functions of $\{\xi\} = \{\xi_1, \xi_2, ..., \xi_m\}$;

– the assembly of the polynomials $\phi_j(\xi)$ in dimension m engenders m-dimensional polynomial chaos $\Psi_j(\{\xi\})$ a polynomial function of the monomials $\xi_1^{p_1}\xi_2^{p_2}...\xi_m^{p_m}$ with $\sum_{k=1}^{m} p_k \leq p$;

– the assembly of the coefficients a_j from the monodimensional basis on the m-dimensional basis defines the coefficients $\tilde{a}_{ij}$;

– response $Y(\{\xi\})$ is therefore expressed as a function of the m-dimensional polynomial chaos $\Psi_j(\{\xi\})$.

Without going into greater detail [LEM 09], we should note that the dimension of the polynomial chaos depends on the order p of the chaos and the number of variables m:

$$J(m,p) = \frac{(m+p)!}{m!\,p!} \qquad [2.5]$$

The J parameters $\tilde{a}_{ij}$ are determined using projection or regression approximation methods [SOI 04, BER 06], and the associated uncertainty may be calculated by resampling [NOT 10].

Polynomial chaos therefore appears as a projection onto a particular orthogonal function basis relating to the probability measure, the choice of which depends on the distribution of the random variable. Hermite chaos is associated with the Gaussian distribution. Thus, two-dimensional Hermite chaos is a response surface with the polynomials $\{1, \xi_1, \xi_2, \xi_1^2 - 1, \xi_1\,\xi_2, \xi_2^2 - 1\}$.

Chaos is useful because it simplifies the calculation of coefficients $\tilde{a}_{ij}$ due to orthogonality. Its limitations lie in relationship [2.5], which increases very quickly as a function of the number of variables and the order of chaos. Chaos of order 1 shows a linear response, order 2 shows a quadratic response and so on. The efficiency of chaos therefore depends on the topology of the response.

A detailed presentation applied to the context of finite elements has been developed by Stefanou [STE 09].

2.3.3. *Reliability analysis – classification methods*

The construction of response Y responds to the requirements of sensitivity analysis, and allows us to deduce the weightings of variables or parameters in the variability of the response (variance analysis, Sobol' indexes [SUD 07a, SAL 08]). As we have seen, the procedure is cumbersome, and reliability analysis only considers the vicinity of failures; we therefore require a criterion for separating acceptable and unacceptable situations. Realization y of Y will belong to the acceptable domain ($y > 0$), also known as the safe domain $\mathcal{D}_s$, or to the failure domain $\mathcal{D}_f$ ($y \leq 0$). Based on a first numerical design of experiments, a small number of deterministic calculations are carried out, including at least one realization from each domain. An iterative procedure is used to enrich the design by carefully selecting additional realizations, each classed according to its sign, based on a probabilistic estimation; in this case, explicit calculations are not strictly necessary, and are neither useful nor economical.

In a probabilistic context, reliability is measured by the notional probability of failure P_f:

$$P_f = \text{Prob}\{Y \in \mathcal{D}_f\}$$

but the classification may also be used with another knowledge model. As before, we are able to obtain responses if the knowledge entity is homogeneous. Two approaches may be used in a probabilistic context: support vector machines (SVMs) and kriging.

2.3.3.1. *Support vector machines*

2.3.3.1.1. Basic concepts

[10]SVMs are an offshoot of Vapnik and Chervonenkis' statistical theory of learning. When using SVMs in reliability analysis, the basic principle involves finding a separator between the domains $\mathcal{D}_s$, containing positive values of a function $Y(\{X\})$, and $\mathcal{D}_f$, containing negative or null values (Figure 2.6). Using all available possibilities (illustration Figure 2.6(a) shows the case of a linear separator), we choose the separator that maximizes the margin between realizations in the two domains (Figure 2.6(b)). The name of the method comes from the vectors supporting the margin (shown by a square in Figure 2.6).

10 From a contribution by J.-M. BOURINET.

Evidently, linear separators generally do not exist in the initial variable space, and a transformation into a space with larger (or even infinite) dimensions is applied (this is always possible).

A transformation of this type is shown in Figure 2.7. A projection operator Φ maps each point $\{x\}$ of the input space onto the corresponding point in the output space. Note that we do not need to identify this operator. The equation of a separator plane in the input space reads (Figure 2.6):

$$\langle\{\varpi\},\{x\}\rangle + b = 0$$

where the product $\langle\{\varpi\},\{x\}\rangle$ is the natural scalar product in the input space. In the output space, the transformation $\{x\} \to \Phi(\{x\})$ substitutes the natural scalar product by the following product in the output space:

$$\langle\{x_1\},\{x_2\}\rangle \to \langle\Phi(\{x_1\}),\Phi(\{x_2\})\rangle = K(\{x_1\},\{x_2\})$$

and we only need to know the kernel function $K(\{x_1\},\{x_2\})$. The choice of a kernel function and the identification of its parameters is therefore a key stage in the SVM method.

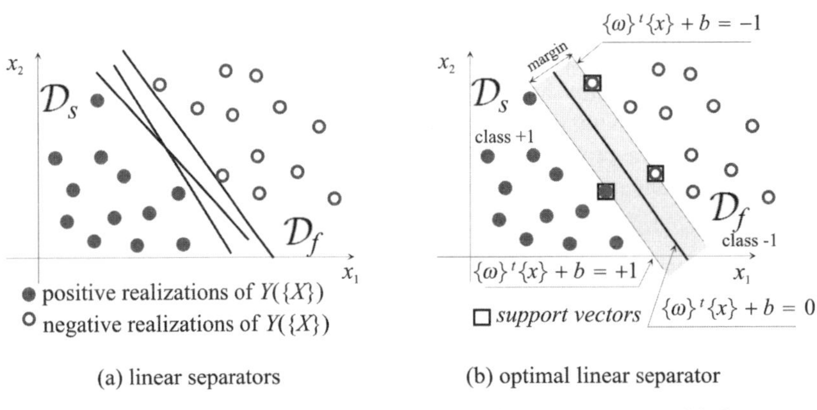

(a) linear separators (b) optimal linear separator

Figure 2.6. *Linear separator between classes of events: success and failure*

The theoretical basis for this tool is described in [BOS 92, VAP 95], and a description of more practical aspects is given in [BUR 98, CRI 00, ABE 10, CAN 13]. SVMs are used in a range of domains, including automatic character recognition, voice recognition, object detection and bioinformatics. The basic focus is the prediction of a quantity of interest $y \in Y(\{x\})$ at a given point

$\{x\}$ based on a known set of learning couples $(\{x\}_1, y_1) \ldots, (\{x\}_N, y_N)$ making up a design of experiments. SVMs were initially introduced for binary classification problems, then used for multiclass problems (Y is defined using a discrete set) then extended to regression [SMO 04] (where Y is defined using a continuous set). SVMs offer the following advantages in relation to other learning techniques:

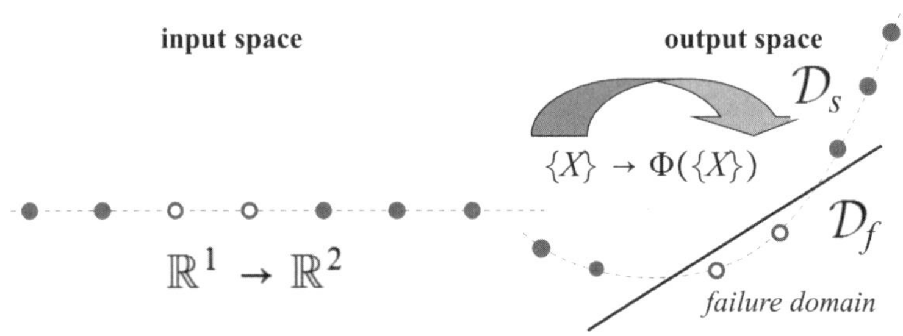

Figure 2.7. *Illustration of linear separation by passing from a space of dimension 1 to a space of dimension 2*

– They produce predictors with good generalization properties (capacity to predict new points $\{x\}$).

– Learning of nonlinear functions is facilitated by the use of kernels, which show proximity relationships between points with relatively little increase in the difficulty of solving the problem, compared to the case of a linear function (the classification problem associated with a nonlinear separator becomes linearly separable in the new space, known as the feature space).

– The optimization problem, which must be solved to find the parameters of the SVM model, is a quadratic programming problem, and the solution of the convex quadratic minimization problem gives us a global optimum. Let us take a set of realizations $\{x_i\}$, $i = 1, \ldots, n$ and a binary classification $c_i = \text{sgn}(y(\{x_i\}))$ for classes $\{+1\,; -1\}$. The margin maximization problem can be written in the form of a minimization problem, as follows:

$$\min \frac{\|\varpi\|^2}{2} \text{ with constraint } c_i\left(\langle\{\varpi\}, \Phi\left(\{x_i\}\right)\rangle + b\right) \geq 1, \; i = 1, \ldots, n \quad [2.6]$$

– The solution obtained using SVM is economical and is generally expressed as a function of a limited number of learning points, or support vectors. This characteristic is particularly helpful when using large-scale learning bases.

2.3.3.1.2. SVM use for evaluating weak failure probabilities

SVMs were first used in reliability analysis in the early 2000s [ROC 02, HUR 04]. The majority of approaches developed aim to treat the reliability problem as a binary classification approach, $\{-1, +1\}$, according to the signs of realizations of $Y(\{x\})$; in this way, we aim to calculate $P_f = \text{Prob}(\text{sgn}\, Y(\{x\}) < 0)$. This type of approach only uses the information provided by the sign of $Y(\{x\})$. Reliability analysis may also be carried out with SVMs used in regression (SVRs), using the value of $Y(\{x\})$ [CHE 07, YU 09, DAI 12]. Reliability analysis using SVMs consists of the stages outlined below.

2.3.3.1.3. Choice of an SVM approach

Classification (SVC) or regression (SVR).

– Choice of the type and kernel of the SVM model

The choice of a kernel type has a major impact in these approaches, and is made *a priori* and independently of the problem in question. To date, reliability studies using SVMs have been based on either polynomial kernels [ROC 02, BAS 10, DAI 12], or Gaussian radials (Gaussian radial basis function (RBF)) [CHE 07, MOS 07, DEH 08, BOU 11], which are expressed as follows:

Gaussian kernel (RBF):

$$K(\{x_1\}, \{x_2\}) = \exp\left(-\frac{1}{2\sigma^2} \left\|\{x_1\} - \{x_2\}\right\|^2\right)$$

where σ is a parameter for which we need to find the optimal value.

– Choice of the regularization parameter

The approximation of a function using a small quantity of available data (few learning points) presents certain challenges. The strict constraint of the optimization problem [2.6] is relaxed by introducing a regularization parameter. Constraint $c_i\left(\langle\{\varpi\}, \Phi\left(\{x_i\}\right)\rangle\ +b\right) \geq 1$ is replaced by $c_i\left(\langle\{\varpi\}, \Phi\left(\{x_i\}\right)\rangle\ +b\right) \geq 1 - \xi_i$, $\xi_i \geq 0$ and the optimization problem becomes (in the case of L1-SVM, by penalizing the sum of the ξ_i):

$$\min \frac{\left\|\varpi\right\|^2}{2} + C \sum_{i=1}^{n} \xi_i$$

with constraints $\xi_i \geq 0$, $c_i\left(\langle\{\varpi\}, \Phi\left(\{x_i\}\right)\rangle + b\right) \geq 1 - \xi_i,\ i = 1, ..., n$

The role of constant $C > 0$ is to create a compromise between the minimization of the classification error in the learning base (an error which corresponds to the empirical risk) and minimizing the complexity of the learning function. The SVM model is then said to have soft margins, and the variables ξ_i are known as slack variables; hard margin models correspond to an infinite value of C.

Up to now, this parameter and the choice of a value have rarely been considered in reliability analysis using SVMs, except in the case of [CHE 07]. This parameter should ideally be selected in the same way as the kernel parameters, something that constitutes a major statistical problem.

– Identification of optimal hyperparameters of the SVM model

The term "hyperparameters" is used to denote the kernel parameter(s) and the regularization parameter C. We need to find optimal values for these hyperparameters in order to guarantee the precision of the constructed SVM model. Cross-validation techniques are generally used in making this choice. This method consists of constructing a number (often large) of SVM models with different sets of hyperparameters, then retaining the model that gives the lowest estimated generalization error via the cross-validation procedure in the learning base.

Different sets of hyperparameters may be generated using a grid search, or more sophisticated techniques (stochastic optimization methods): simulated annealing [PAI 06, LIN 08a], genetic algorithms [PAI 06, CHE 07] or particle swarms [LIN 08b].

Simpler techniques have also been used in the past: lowest degree (separator is as general as possible), which leads to error-free classification of learning points in the case of a polynomial kernel [BAS 10], or highest bandwidth σ, which gives a maximum learning error of 2% on learning points in the case of a Gaussian RBF [MOS 07].

– Adaptive designs for calculating weak probabilities

The use of a learning base chosen in advance does not generally allow us to construct an SVM model, which is sufficiently precise for weak failure probabilities. The approaches used consist of constructing a set of SVM models sequentially, using a learning base that is enriched over time with the addition of new points (chosen based on the constructed SVM), leading to the creation of a final SVM that gives the closest possible approximation of the real limit-state of the system in question [MOS 07, DEH 08, YUA 09, BAS 10, BOU 11].

Different authors have used different methods to obtain learning points: Latin hypercube designs [MOS 07], centroidal Voronoi tessellation (CVT) designs [BAS 10] or random generation based on Markov chains [YUA 09, BOU 11, DAI 12]. Criteria are generally applied in order to select a limited number of learning points, for example the generated points that fall closest to the constructed SVM separator [MOS 07, BOU 11], points that change classes in two consecutive constructions of the SVM separator [BOU 11], or a selection based on the distance between points (points on the margin, points from zones with the lowest density of learning points) [BAS 10] □.

Age		FORM	Subset	Subset	^{2}SMART
10 years	N calls	200	16,000	80,000	4,500
	P_f (10^{-8})	4960	3.36	1.89	1.96
	β	3.89	5.40	5.50	5.49
20 years	N calls	200	16,000	80,000	4,500
	P_f (10^{-8})	170,000	6.66	8.71	10.23
	β	2.93	5.27	5.22	5.20
40 years	N calls	200	14,000	70,000	4,000
	P_f (10^{-7})	111	4.75	4.15	4.02
	β	4.24	4.90	4.93	4.93
60 years	N calls	200	14,000	70,000	4,000
	P_f (10^{-7})	10,600	6.70	8.47	10.07
	β	3.07	4.83	4.79	4.75

Table 2.6. *Reliability-based results obtained for the PROSIR project with a random default*

2.3.3.1.4. Illustration

In his doctoral thesis, Deheeger [DEH 08b] compares the efficiency of the ^{2}SMART (subset simulations by support-vector margin algorithm for reliability estimation) method [DEH 07, BOU 11], which associates SVM and subset simulation, for both academic and industrial cases. The PROSIR project [COL 06, DEH 08] consisted of testing different methods for a benchmark of crack propagation in a reactor vessel. The behavior model defined a highly complex failure domain with very weak probabilities. The data model included eight random variables, with parameters defined in a coherent manner, bearing no relationship to the real data. Table 2.6 gives an overview of the results as a function of time. FORM is a linear approximation, which is economical but offers no possibility of validation and does not produce satisfactory values. The two subset calculations are considered as reference points, but involve a high number of calls of the behavior model.

The reliability indexes $\beta = -\Phi^{-1}(P_f)$ (the right measure scale) obtained by ^{2}SMART are very close to the reference values, and only required $4,500$ calls, compared to $80,000$ for the reference values. This demonstrates the efficiency of the method; a measure of the confidence interval is now needed.

2.3.3.1.5. Conclusion

Classification methods contribute to relatively economical implementation of reliability calculations, but significant gains in terms of technical applications are not always observed in practice. As for many other approaches, an important issue remains concerning validation in cases where it is not possible to compare results with those of a Monte-Carlo simulation. In this section, SVMs have been presented in a probabilistic context, but classification may also be carried out with sets of realizations generated using a different type of model. This may be a path to follow for the implementation of other knowledge models.

2.3.3.2. *Kriging methods*

2.3.3.2.1. Principle

[11]Kriging, developed by Krige [KRI 51] and theorized by Matheron [MAT 73], is used in the context of reliability and considers the performance function $Y(\{x\})$ as a realization of a Gaussian process noted $\mathcal{Y}(\{x\})$. The first stage of kriging consists of fixing the parameters of this process based on a numerical design of realizations $Y(\{x\}^{(i)})$. The best linear unbiased predictor (BLUP) is then used to predict the unknown value of the performance function for any non-observed realization of $\{x\}$. In the context of reliability, we only need to predict the degree of membership of $Y(\{x\} \neq \{x\}^{(i)})$ to $\mathcal{D}_f$.

Kriging presents the following two key characteristics, which are useful in reliability analysis and optimization.

– *Interpolation*: the kriging average $\mu_{\tilde{Y}}(\{x\})$, used as an estimator of $Y(\{x\})$, passes exactly through the points of the numerical design of experiments $\{x\}^{(i)}$, i.e. $\mu_{\tilde{Y}}(\{x\}^{(i)}) = Y(\{x\}^{(i)})$ for any $\{x\}^{(i)}$.

– *Use of the estimation variance as an a priori indicator of prediction uncertainty*: for a non-observed point, kriging provides not only a prediction of the performance function, but also a measure of uncertainty for this prediction in the form of the kriging variance. Iterative enrichment of the design of

11 From contributions by N. GAYTON and V. DUBOURG.

experiments based on this variance is a helpful strategy for reducing the number of mechanical calculations; this approach is generally known as active learning.

2.3.3.2.2. Use of kriging to evaluate failure probabilities

To the best of our knowledge, [ROM 04] was the first reference to recommend kriging for reliability analysis. In this study, kriging was used in the same way as polynomial response surfaces, i.e. in an iterative manner and without taking account of the kriging variance in order to enrich the design of experiments. Kaymaz [KAY 05] used kriging in a similar way. Since then, kriging has been used in active learning procedures [BIC 08, RAN 08, PIC 10, BEC 12, DUB 11a, ECH 11] where the numerical design of experiments is enriched based on the evaluation of a learning function. The general procedure is as follows:

1) Construct a kriging meta-model of Y with an initial design of experiments.

2) Find the point(s) which maximize the learning function.

3) Enrich the design of experiments with one or more points, calculating the associated function Y.

4) Update the kriging meta-model and repeat the process from step 2.

The learning function should be defined in accordance with the problem under consideration (optimization, reliability, etc.). In reliability analysis, only points close to the limit-state $Y(\{x\}) = 0$ are of interest. At any point within the design of experiments, the prediction $Y(\{x\}) \in D_f$ is certain; outside of this design, the prediction is only probable.

2.3.3.2.3. Learning functions

A number of different learning functions have been proposed in the context of reliability analysis. In this section, we will limit our presentation to the three learning functions that we consider to be the most efficient.

– The EFF function

The learning process proposed by the efficient global reliability analysis (EGRA) [BIC 08] method is based on a learning function known as the expected feasibility function (EFF). Inspired by the work of Ranjan *et al.* [RAN 08], the EFF expresses the expectation that the performance function at a non-evaluated point $\{x\}$ will respect the constraint $Y(\{x\}) \in [-\epsilon(\{x\}); +\epsilon(\{x\})]$. This criterion allows us to specify the extent of the imprecision of the kriging meta-model for limit-state prediction, and is defined proportionally in relation to the kriging standard deviation. The expression reads:

$$\mathrm{EFF}(\{x\}) = \mathrm{E}\left[\epsilon(\{x\}) - \min(|\tilde{Y}(\{x\})|, \epsilon(\{x\}))\right]$$

High EFF values are obtained for points with a kriging average close to 0 or for points with high kriging variance. The performance function is evaluated iteratively at the points where the EFF function is greatest.

– The U function

The U learning function, proposed in [ECH 11], is an adaptation of a criterion proposed in [KUS 64] and [COX 97] (lcb optimization criterion). The U criterion indicates the distance between the kriging average and the threshold considered for function Y in terms of a number of kriging standard deviations:

$$U(\{x\}) = \frac{|\mu_{\tilde{Y}}(\{x\})|}{\sigma_{\tilde{Y}}(\{x\})} \qquad [2.7]$$

Criterion U is linked to the probability of misclassification of point $\{x\}$. As for EFF, it promotes enrichment at points in the initial design where the kriging prediction is close to 0 and/or the estimation variance is high.

– The $\pi \times \phi_n$ function

The learning function used by [DUB 11b] aims to enrich the initial design at points in the vicinity of the limit-state in the most probable zones. This compromise is obtained by multiplying a "probabilistic classification" function π by the probability density function of the vector $\{x\}$ (the vector of the normal standard distribution in dimension n, denoted as ϕ_n). The probabilistic classification function expresses the probability that the limit-state function will have a negative value at a given point $\{x\}$ in relation to the kriging uncertainty (Gaussian):

$$\pi(\{x\}) = \Phi\left(\frac{-\mu_{\tilde{Y}}(\{x\})}{\sigma_{\tilde{Y}}(\{x\})}\right) \qquad [2.8]$$

where Φ is the function of the normal standard distribution. Unlike the criteria discussed previously, this criterion may be used to refine the precision of the limit-state exclusively in the most probable zones (which make a significant contribution to the probability of failure). In AK methods [ECH 11], this property is obtained by a specific use of the learning function U. The function is evaluated using a single sample of the vector $\{x\}$, used throughout the algorithm up until the final estimation of the failure probability.

2.3.3.2.4. Searching for the optimum learning function

Learning functions are highly multimodal, and finding an optimum is a complex matter. The EGRA method uses the gradient-free DIRECT (dividing rectangles) method [JON 93, GAB 01] to search for a global optimum. This method consists of iterative divisions of the search space into hyper-rectangles. The EFF function is calculated at the center of these hyper-rectangles. While the evaluation of learning functions does not require us to calculate the performance function, its inclusion in an optimization method can be expensive in terms of calculation time, and it does not guarantee convergence to the global minimum. The classification principle proposed in AK methods [ECH 11] avoids searching for this optimum in a continuous space, and the algorithm focuses on a Monte-Carlo type sample. The search for the optimum of the learning criterion is then carried out in a discrete manner using this finite population, making the operation considerably easier.

Learning method [2.8] makes use of the fact that the high level of multimodality of learning functions means that there is not really one single better point to add to a given iteration of the learning algorithm. The authors of the method therefore aim to add several points to each iteration. These points, situated in the modes of the $\pi \times \phi_n$ learning function, are identified by a sampling-clustering technique, based on work presented in [DEH 08a, BOU 11] on SVMs. This technique makes use of the availability of high-performance computation (HPC) materials (e.g. PC clusters), and the number of points added for each iteration is determined according to the resources available (processors, RAM, licenses, etc.).

2.3.3.2.5. Quantification of the substitution error

The vast majority of reliability methods using kriging estimate the probability of failure by substituting the final (refined) kriging meta-model for the original limit-state. This approach raises questions concerning the quality of the approximations. Methods based on the continuous optimization of learning functions stop enriching the design of experiments once the value of the learning function at the obtained optimum is below a certain threshold. However, in certain cases, the global optimization algorithm may not converge to the true global optimum, missing a region of the space where the learning error may be significant. AK methods [ECH 11] avoid this source of error by working with the population used to estimate the probability of failure from the outset, and ensuring that the learning function is below a certain threshold for all individuals in this population.

The Meta-IS method [DUB 13] does not use substitution. The final kriging meta-model is used to construct a quasi-optimal instrumental density (proportional to the $\pi \times \phi_n$ learning function presented earlier), which allows us to estimate the probability of failure on the original limit-state using preferential sampling [RUB 08]. Although this method requires further calls to the original limit-state function in cases where other methods use substitution, the number of additional calls is minimized due to the quasi-optimality of the instrumental density. In practice, in cases where substitution could have been used, this validation only requires tens of new calls. As the precision expected from kriging declines, however (e.g. as the number n of random variables increases), this method gives a true estimation of the probability of failure, the quality of which is measured by the estimation variance alone, as with other similar reduced variance simulation methods □.

2.3.3.2.6. Conclusion

Several research teams are currently working in kriging, which has already proved its efficiency and is one of the most promising reliability calculation methods. Unlike other methods, it does not call for large quantities of calculations with no possibility of identifying which are most relevant in relation to the objective beforehand. In cases where n evaluations of a performance function are available, kriging provides a way of choosing the next m evaluations needed to validate the desired result, in this case the probability. Note, however, that the cost increases as a function of the number of variables.

Kriging has already proved useful in responding to industrial questions involving very weak probabilities. Adaptive implementation for reliability analysis has produced good results in aeronautics, limiting the number of calls of numerical models, which are expensive in terms of calculating time [ECH 13]. In the same way, very weak probabilities have been identified in the case of reactor vessels, using a particularly conservative model [DUB 12]. Finally, kriging has also contributed to the development of reliability-based design optimization (RBDO) methods for optimal, reliable design [DUB 11a].

2.3.4. *Model reductions*

[12]All calculations concerning uncertainty propagation are based on a behavior model, using a carefully reiterated deterministic algorithm. The most

12 From a contribution by J.-Y. DULONG.

complex models currently require too much processing power to be used directly. These models can be reduced by constructing a response using a simple basis in a defined domain of validity. They may also be reduced by using an algorithm to extract their main properties. This forms the basis of model reduction techniques.

The term "model reduction" is generic and covers a range of methods used in different domains; the plural term "model reductions" is therefore preferable. These methods all have a shared aim of reducing the time needed to calculate the solution to a given physical behavior model, i.e. the variable of interest Y, without excessive reductions in the desired precision level.

The approach used for model reductions is independent of consideration of the uncertainty associated with the model: the reduced behavior model rapidly calculates the instance associated with the variable of interest $Y = \mathbf{M}(\{X\})$ for a given instance of random variables (realization) $\{X\}$. Model reductions are therefore unconnected with stochastic hypotheses. Consequently, they may be used with most of the uncertainty calculation methods presented elsewhere in this book.

2.3.4.1. *Strict minimum reduction*

The first type of model reduction is "strict minimum" modeling, based on engineering knowledge. For instance, it is not necessary to create a finite element model in cases where an analytical approach, for example of the strength of materials type, would suffice. Finite element modeling is used as a point of reference in the mechanics industry due to its capacities and generic character. The problem with this approach is that it has progressively replaced all other methods in the domain, despite its drawbacks, notably the time required to prepare and solve a model. In cases where finite element methods are unavoidable, the careful choice of mesh density and element types (beams or shells rather than volumes, for example) allows reductions in calculation time. These model simplification methods operate by using as much physical knowledge as possible in analytical form rather than using numerical approximations. This approach makes the model more specific, but also increases its speed, and the models produced are often sufficient: models do not need to produce more information than is relevant for the required decisions.

2.3.4.2. *Reduction by learning using response surfaces*

Digital or experimental results can be used to construct a response surface using a function basis. Polynomial chaos (section 2.1.4) is the stochastic version of polynomial responses. These responses have been the subject of extensive research [FAR 89, EL 91, RAJ 93, ENE 94b, KIM 97], with each team adding new elements; however, none of these projects gives sufficient consideration to validation [GAY 02].

Responses of this type are often considered to be too "smooth", and attempts have been made to find richer bases: spline functions, Bezier surfaces [BAR 06] and neural networks [LEM 00, IPV 06]. The key issue in this case is response validation, which is carried out by separating the numerical design of experiments into two subsets. The first subset is used for learning and the second for validation. In this context, neural network theory gives the best compromise between learning and the capacity for generalization. A response may be constructed in a physical space or in a normed random space; the second method is preferable, as it allows us to obtain a measure of the distance between points in the design.

2.3.4.3. *Algorithmic reduction*

The third type of model reduction is based on mathematical and numerical techniques; instead of attempting to simplify the physics involved in a model, we look for a different way of expressing the model in order to obtain a more economical numerical solution. The models used in these cases are always discretized (in time and space) to enable numerical resolution. The problem then consists of solving a series of systems of equations as quickly as possible.

2.3.4.3.1. Control theory

Model reduction was initially developed within the context of control theory, and was generally limited to systems of ordinary differential equations modeling a real dynamic system. In such cases, this real system was controlled in real time by a number of actuators, themselves controlled by the solution to the system of equations. In cases involving a bulky system of equations, where the required solution time was greater than the response time of the real system, a reduced model, quicker to solve but preserving the behavior of the system (responses, stability, etc.) was used. However, to the best of our knowledge, these studies did not take account of uncertainty. We will not go into greater detail on the subject here; a thorough introduction has been provided by Schilders *et al.* [SCH 08].

2.3.4.4. *Reduction of the discretization basis*

2.3.4.4.1. Principle

More recently, model reduction has undergone developments in computational mechanics, where it is used for structural calculations, and in fluid mechanics. In both cases, problems are almost always formulated using finite element or finite difference spatial and temporal discretization, producing a series of systems of equations. Let ξ be a drawing variable of an instance of the data $\xi \longrightarrow X(\xi)$. The vector $q(\xi)$, which is the discrete form of the continuous field $u(x, \xi)$, is the solution to a nonlinear matrix system with n degrees of freedom:

$$K(X(\xi), q(\xi))\, q(\xi) = F(X(\xi)) \qquad [2.9]$$

The variables of interest $Y(\xi) = \mathbf{M}(X(\xi))$ are then processed based on the solution field $q(\xi)$. The specific case of linear behavior where loading alone depends on the random aspect presents no difficulties, as matrix K, which is constant, can be definitively inverted (or triangularized):

$$q(\xi) = K^{-1}(X(\xi))\, F(X(\xi))$$

Next, we must simply generate instances $\xi \longrightarrow F(X(\xi))$ to obtain a rapid solution to $q(\xi)$.

In the case of nonlinear problems, $K(X(\xi), q(\xi))$, each instance of $\xi \longrightarrow X(\xi)$ must be solved in full: in an industrial context, the calculation time involved rapidly becomes problematic. To reduce the size of the system of equations, the continuous solution $u(x, \xi)$ is projected onto a basis Φ of dimension $n \times m$, containing only m modes $\phi_i, i = 1, ..., m$ ($m \ll n$), which are representative of the solution. This equates to separating the space variables x from the drawing variables ξ:

$$u(x, \xi) = \sum_{i=1}^{m} \phi_i(x)\, a_i(\xi) \text{ or after discretization:} \{q(\xi)\} = [\Phi]\ \{a(\xi)\} \qquad [2.10]$$

or, in more simple terms: $q(\xi) = \Phi\, a(\xi)$, where the vectorial dimension is not expressed in writing.

If the basis Φ is known, the system of equations [2.9] may be solved using a Galerkin type projection:

$$\Phi^t\, K(\xi)\, \Phi\, a(\xi) = \Phi^t\, F(\xi) \qquad [2.11]$$

This produces a reduced system [2.11], which allows us to calculate the amplitudes $a_i(\xi)$ of the modes $\phi_i(x)$ using a square reduced matrix $K_r(\xi) = \Phi^t\, K(\xi)\, \Phi$ of dimension m, and a second reduced component $F_r(\xi) = \Phi^t\, F(\xi)$. Using this approach, we move from a series of $n-$dimensional equation systems, which may attain several million degrees of freedom, to a series of reduced systems of dimension m, with, at most, a few tens of modes. The heavier the initial model, the greater the benefits, which may be obtained using this method.

2.3.4.4.2. Determination of the basis Φ

The difficulty of this approach resides in the identification of a basis Φ, which is sufficiently rich to describe all solutions $q(\xi)$. The first proposal to tackle this issue [KRY 01] consists of carrying out several non-reduced calculations for several instances of $X(\xi)$.

The associated solutions, known as snapshots, are considered to describe the set of possible solutions, whatever the future instances may be. The basis Φ is extracted from the correlation matrix of the set of these solutions. This method, known as proper orthogonal decomposition (POD), is a singular value decomposition (SVD) of the correlation matrix. Thus, the vectors of the basis Φ are decorrelated. This approach is identical to that used in the Karhunen–Loève transform (KLT), discussed previously, to decorrelate data fields (see 2.3.2.4). However, in the case of POD, the autocorrelation function of the solution field is unknown: an approximation must therefore be established through initial precalculations.

This approach was then applied to stochastic contexts [MAT 05, DOO 07], but the relevance of the basis Φ is strongly linked to the choice of instances in the precalculation phase; it is not currently possible to precisely assess this relevance. For the same reasons, the basis Φ is able to correctly describe solutions that are close to those obtained in the precalculation phase, which fall into Mediocristan. For results are that further from those of the precalculation phase, we do not know if the obtained description is correct; this approach is therefore unsuited to the exploration of Extremistan.

Once again, we must choose an initial design of experiments, then enrich it in order to cover a sufficient domain of generalization.

Certain authors [RYC 05] have explored the possibility of an *a priori* approach, constructing the projection basis Φ as the series of calculation advances, in order to tackle the lack of confidence inherent in a basis calculated *a posteriori*. This approach consists of solving the reduced equation [2.11], deducing the approximate solution [2.10] using the current basis Φ and verifying the residual of the non-reduced equation [2.9] in order to verify the quality of the approximate solution. If the residual is too high, then the basis is not sufficiently rich, and is enriched using the residual r:

$$\Phi = \{\Phi, r\}$$

The new, enriched basis is reused to continue the solution process. This procedure allows us to progressively construct a basis Φ of the Krylov space describing the solution, in the case of a linear problem.

The most recent approach developed for constructing the projection basis Φ is known as proper generalized decomposition (PGD) [AMM 06, NOU 10]. Unlike previous approaches, PGD involves the simultaneous construction of the basis Φ and the reduced solution $a(\xi)$, rather than estimating a basis Φ and using it to construct a reduced response using equation [2.11]. This approach was developed in the context of a transitory deterministic problem, proposing a separation of the space-time variables of the solution $u(x,t)$, analogous to [2.10]. The procedure presumes that the first m modes and their amplitudes are known for the whole space and time (m is zero at the point of initialization): the problem consists of finding the couple $\{\phi_{m+1}(x), a_{m+1}(t)\}$:

$$u(x,t) = \sum_{i=1}^{m} \phi_i(x)\, a_i(t) + \phi_{m+1}(x)\, a_{m+1}(t)$$

Thus, we find the solution $u(x,t)$, in the form of separated variables, by successive enrichments of $\{\phi_{m+1}(x), a_{m+1}(t)\}$ couples. Two systems of equations are extracted from the formulation in order to identify these couples. The first supposes that $a_{m+1}(t)$ is known, and allows us to calculate $\phi_{m+1}(x)$. The second uses knowledge of $\phi_{m+1}(x)$ to calculate $a_{m+1}(t)$. The two systems are solved by iterations until they converge. At a fixed point, this procedure gives the solution to the couple $\{\phi_{m+1}(x), a_{m+1}(t)\}$.

In the case of a separation of the space-time variables, the major advantage of PGD lies in the simultaneous solution of a spatial mode $\phi_{m+1}(x)$ and its amplitude $a_{m+1}(t)$ for the whole time interval. The solution is not incremental over time, something which considerably reduces calculation time: an approximation $\{\phi_1(x), a_1(t)\}$ of the global solution is obtained at the first fixed point. This approach, used for deterministic solutions, has been adapted for stochastic contexts by Nouy [NOU 08].

Note that all of these model reduction methods using a projection of the solution onto a basis Φ are highly intrusive, in that they require modification of the core of the calculation program, i.e. its resolution kernel $\square$.

2.3.5. *Quantifying uncertainty*

Given a data model and a behavior model, both of which include uncertain parameters represented by bodies of knowledge $\{P\}$ and $\{X\}$ as noted in relationship [2.2]:

$$\{Y\} = \mathbf{M}\Bigg(\underset{\text{behavior model}}{(\mathbf{F}, \{P\})}\,,\ \underset{\text{data model}}{\{X\}}\Bigg)$$

and a measure μ associated with a component Y of $\{Y\}$ for which an acceptability ε is accorded at a threshold a, i.e.:

$$\mu\left[Y\left((\mathbf{F}, \{P\}), \{X\}\right) \geq a\right] = \varepsilon$$

we need to identify the limits:

$$\inf_{(\mathbf{F},\mu)\in\mathcal{A}} \mu\left[Y\left((\mathbf{F}, \{P\}), \{X\}\right) \geq a\right] \leq \mu\left[Y\left((\mathbf{F}, \{P\}), \{X\}\right) \geq a\right] \quad [2.12]$$

$$\mu\left[Y\left((\mathbf{F}, \{P\}), \{X\}\right) \geq a\right] \leq \sup_{(\mathbf{F},\mu)\in\mathcal{A}} \mu\left[Y\left((\mathbf{F}, \{P\}), \{X\}\right) \geq a\right]$$

for any admissible $(\mathbf{F}, \mu)$ belonging to a set $\mathcal{A}$, generally of infinite dimensions, of functions $\mathbf{F}$ and measures μ. This UQ problem [2.12] needs to be processed for each component of $\{Y\}$. We are then faced with the issue of aggregating these limits, which may be dependent.

The approaches proposed in the previous sections may be considered as attempts at partial resolution of [2.12] under more or less explicit hypotheses relating to the knowledge entity (deterministic, possibilistic, probabilistic, etc.) of data $\{X\}$ and parameters $\{P\}$ and to the behavior description $\mathbf{F}$ (functions, calculation algorithms, approximations, etc.).

The optimal uncertainty quantification (OUQ) approach, described in detail in [OWH 13], demonstrates the approximation of a solution to [2.12] when the measure is a probability. The authors use a range of tools to reduce the dimensions of set $\mathcal{A}$: expert analysis of $\mathbf{F}$, $\mathcal{A} \to \mathcal{A}_{\mathbf{F}}$; Markov concentration inequalities (a generalization of the Bienaymé–Chebyshev inequality) $\mathcal{A} \to \mathcal{A}_{Mrkv}$; McDiarmid's inequality [MCD 98] $\mathcal{A} \to \mathcal{A}_{McD}$; reduction theorems with convexity hypotheses, etc., then a differential evolution optimization algorithm [PRI 05].

A first optimization using the available information provides a first set of limits, and new theoretical or experimental information can be carefully added in order to reduce the limits.

This principle can be illustrated using a simple example [OWH 13]. Let us suppose that [2.2] is written as $Y = \mathbf{M}(X)$, where Y is a random variable with real values in the interval $[0,1]$ and $a \in (0,1)$. We wish to find which probability measures ν of $\mathbb{R}$ are admissible for the probability distribution of Y. Without using additional information, the set $\mathcal{A}$ is:

$$\mathcal{A} = \{\nu \mid \nu \text{ is a probability measure on } [0,1]\}$$

and the optimal upper limit is trivial:

$$\text{Prob}\,[Y \geq a] \leq \sup_{\nu \in \mathcal{A}}\,[Y \geq a] = 1$$

Let us now add additional information: the mean of Y is equal to $m \in (0,a)$. Set $\mathcal{A}$ is thus:

$$\mathcal{A}_{Mrkv} = \left\{ \nu \,\middle|\, \begin{array}{c} \nu \text{ is a probability measure on } [0,1] \\ \text{E}_\nu\,[Y] = m \end{array} \right\}$$

and the Markov inequality gives:

$$\text{Prob}\,[Y \geq a] \leq \frac{m}{a}$$

The concentration inequalities concern the distribution around the mean. The McDiarmid inequality applies to m independent random variables $\{X\}$ and a function $\mathbf{F}$ verifying the hypothesis of bounded differences:

$\forall i \in [1,m]\,,\, \exists\, c_i(\mathbf{F}) > 0,$

$$\sup_{(x_1,\ldots,x_m)} \sup_{x_i'} |\mathbf{F}(x_1,\ldots,x_m) - \mathbf{F}(x_1,..,x_{i-1},x_i',x_{i+1},\ldots,x_m| \leq c_i$$

c_i therefore represents the maximum amplitude of the behavior model $\mathbf{F}$ for factor i. The inequality reads:

$$\mu\left[\mathbf{F}\left(X_1, ..., X_m\right) - \mathrm{E}_\mu\left[\mathbf{F}\right] \geq a\right] \leq \exp\left(-\frac{2a^2}{\sum_{i=1}^m c_i^2(\mathbf{F})}\right)$$

These bounds explore the vicinity of the mean, and therefore fall into Mediocristan. Are they still meaningful in Extremistan? As the authors have highlighted, the OUQ approach is an as-yet unfinished attempt to provide a universal response to the question of quantifying uncertainty, in the same way as Kolmogorov provided a universal response for probability theory.

2.3.6. *Conclusion*

A great deal of work has been carried out in cases where the knowledge entity constitutes a homogeneous model, particularly in the probabilistic domain. Efficient procedures are now available for both sensitivity and reliability analysis, but still require significant calculation resources. Work is still in progress, and the development of computer clusters will provide us with new possibilities.

However, for cases where the knowledge entity constitutes a heterogeneous model, i.e. where the level of knowledge of each variable is not the same, homogenization procedures need to be defined before transmission calculations can be carried out. This issue may be resolved within the formal context of OUQ, attempting to bring together all of the approaches discussed in this chapter.

Once measures have been obtained for variables of interest, these measures need to be aggregated to create a decision process in order to achieve the best possible compromises.

Chapter 3

Decision Support under Uncertainty

3.1. Decision support in design

3.1.1. *Decision support*

[1]In the previous chapters, we focused on the intrinsic uncertainty of variables and their propagation. The design cycle is often represented as shown in Figure 3.1. The conceptual design phase is followed by the embodiment design phase, which involves choices with a decisive impact on the subsequent demonstration of robustness and reliability; this phase precedes the detailed design phase, which benefits from the use of uncertainty analysis methods. The uncertainty involved in the embodiment design phase is linked to the decisions made by the designer, rather than the physical aspects; these decisions need to be robust (robust decision-making). A wide variety of different options are available, and pointers are needed to help us in making the best choices.

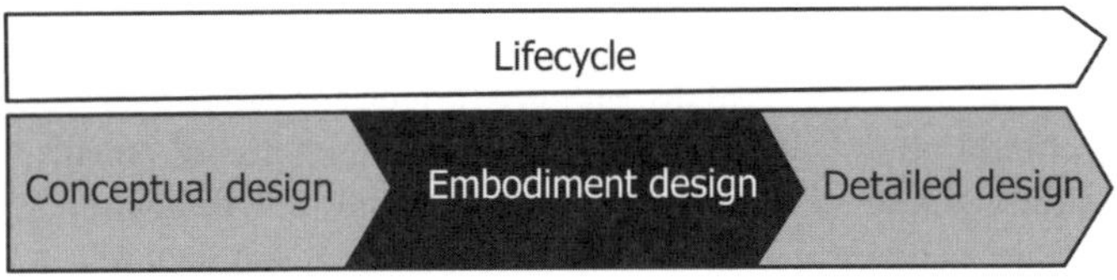

Figure 3.1. *The first phases of the design cycle*

1 From a contribution by S. LEDOUX and P. SÉBASTIAN.

The scientific domain of *decision support* was initially developed in relation to game theory, and it made significant progress in connection with the *sciences of the artificial* [SIM 69]. Uncertainty arises from the object – subject (real object – idealized object) duality; this duality is undecidable, and it can be resolved only partially by using an iterative process. As the design process creates artificial objects, it is subject to the imperatives of procedural or limited rationality [SIM 82]. The solution-seeking process is based on the use of heuristics, i.e. flexible and adaptive design rules that tend to guide research toward preferable solutions, while simultaneously considering a wide variety of aspects of product design. These rules have a limited sphere of usage and develop as the result of experience (expertise). Design support may consist of either creating "artificial designers", capable of recreating the intelligence of designers using calculations (search heuristics, rules, etc.), or helping designers to model their subjective preferences using an algorithmic *decision support* system. The sole aim of a *design support* system is to guide designers toward preferable solutions; these systems do not, *a priori*, aim to question or reconstruct the rules used by the designers.

The difficulty of this approach lies in the definition of the invariants connected with design problems, which go beyond the subjective views of the designer. These elements, present in all design problems, are essential for *decision support* systems, which use invariants to enlarge their domain of usability. The main design invariants are the following:

– the notion of priority, structured by product functions (services, constraints, etc.);

– the relative comparison of values, which forms the basis for the rules used by designers when seeking solutions;

– certain classes of variables that support designer decisions;

– the heuristic approach that consists of comparing, selecting and eliminating values and solutions.

These invariants are used to produce preferable solutions based on a multiple-criteria optimization process, guiding the choice of physical values for pre-dimensioning, standard components, system architecture, materials and the main geometric dimensions. The selected architectural solutions, chosen from a set of possible solutions meeting structural and functional criteria, must satisfy the preference criteria and optimized objectives. A compromise is always involved to ensure the availability and security in connection with planned maintenance and logistical support.

3.1.2. *Modeling decision support*

3.1.2.1. *Behavior models*

To model designer preferences, we first need to structure the variables involved. We must therefore identify the design variables $\{X\}$ that characterize the product or system in question. These variables are exogenous, i.e. external to the model implemented by the designer to observe the behavior of the product or system. Their values must be determined using expertise or feedback from previous experiences, with the help of an optimization code. Using other endogenous variables, which are known as performance (or observation) variables, denoted by $\{Y\}$, we observe the performances resulting from the choice of design variables. The values of the performance variables are thus a result of the behavior model $\mathbf{M}$, following relationship [2.2]:

$$\{Y\} = \mathbf{M}((\mathbf{F}, \{P\}), \{X\})$$

This relationship also shows the behavioral uncertainty $(\mathbf{F}, \{P\})$. Model $\mathbf{M}$ is an observation model; it is of a physical and/or economic type and reflects the effective behavior of a product. At this stage, designer preferences are not expressed in the model, but the existing model is used to calculate the performance variables, which will be used to support the product performance indicators in the preference model.

3.1.2.2. *Preference models*

The preference model is based on a phase in which performance variables are interpreted, i.e. transformed into performance indicators, denoted as $\{Z\}$. In cases with multiple performance indicators, a unique global preference indicator is used, denoted as Z_G and based on an aggregation operation. Thus:

$$\{Z\} = \delta(\{Y\}) \text{ and } Z_G = \xi(\{Z\})$$

where functions δ and ξ are interpretation and aggregation models, expressing the satisfaction associated with each value of the physical variables $\{Y\}$ (the value of the variable is interpreted in a qualitative manner) and with a global satisfaction value, which aggregates the values of $\{Z\}$, weighting them according to priority levels, respectively (Figure 3.2).

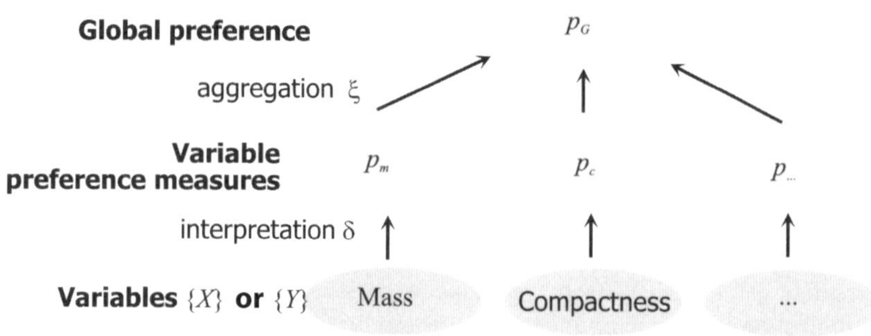

Figure 3.2. *Interpretation and aggregation of variables in a preference model*

The interpretation operation consists of transforming a variable defined in a physical space (dimensions, mechanical constraints, temperatures, costs, etc.) into a variable defined in a metric design space [ANT 05, OTT 93], measuring a degree of satisfaction. The design metric is used to transform the designer's existing subjective vision. It allows us to use both physical and economic variables, with fundamentally different natures, in the same measurement space. Thus, these variables become comparable, which is necessary because the designer aims to create a product that both provides a service (the service function of the product) and satisfies a set of constraints (technical and economic functions). This is the case, for example, of the economic constraint associated with the cost of a product, which must not exceed a fixed threshold. The idealized product is defined using the parameters of the interpretation function, and the priority order of functions (service functions or product constraints) is defined by the parameters of the aggregation function [DER 80, JEO 09, SCO 98].

3.1.2.3. *Interpretation and aggregation functions*

By defining the parameters of functions δ and ξ, the designer produces a mathematical formula of an algorithmic specification, using a set of flexible satisfaction levels associated with priority levels. The preference measure requires us to have a continuous or discrete set of solutions and an order relationship between these solutions: utility, acceptability or desirability functions [HAR 65, DER 80].

Each element X of the set is given a value $p_X \in [0, 1]$, as shown in Figure 3.3. In terms of objectives, we may aim either for a target threshold or for a limitation. The acceptable constraint zone provides the degree of freedom required to avoid excessive reduction of the solution space. Smooth mathematical forms are proposed. For each variable, we thus obtain a classification of solutions, with satisfaction levels defined as acceptable or unacceptable.

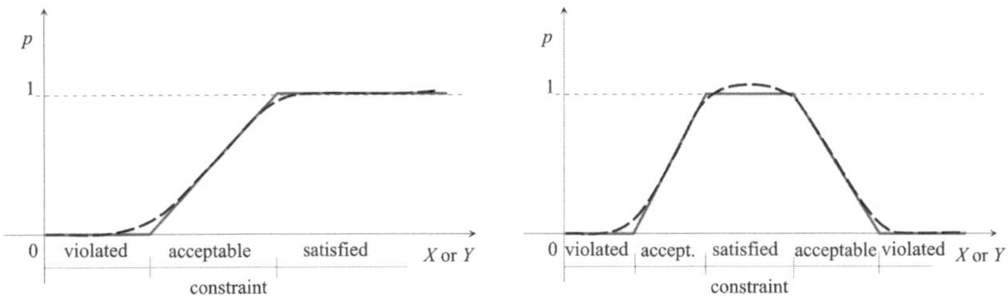

Figure 3.3. *Typical parameterizing of a preference function with a limitation or target objective*

Aggregation is based on a combination of preferences, according to different possible models:

– product of the preferences $p_G = \prod p_i(y_i)^{w_i}$ [DER 80];

– minimum rule $p_G = \min(p_i(y_i))$ [KIM 00];

– general weighted average $p_G = \left(\frac{\sum w_i\, p_i^s}{\sum w_i}\right)^{\frac{1}{s}}$ [SCO 98].

where w_i are weightings equivalent to criticality levels (product of the occurrence, the incidence and the detectability of failure) of service or constraint functions, and s is a parameter that varies from ∞ to 0.

In terms of industrial developments in the design process, this approach has many advantages and drawbacks. The mathematical formalism speeds up the numerical treatment of information, but is only fully efficient if the processing chain is sufficiently complete, robust and reliable. The development cost involved in parameterizing the preference model is balanced out by the capitalization of tacit or poorly formalized knowledge available within the company. In certain companies, the formalization of preferences may be problematic due to confidentiality or conflict issues, both within and outside the company. Finally, if the preference model is expressive and gives a faithful reproduction of designer objectives, this approach provides a framework for the scientific construction of company-specific design strategies.

3.1.3. *Multi-criteria decision analysis (MCDA)*

3.1.3.1. *Optimization problems*

Preference interpretation and aggregation form the basis of *multi-criteria decision analysis* (MCDA), a scientific domain with strong links to optimization theory [GRE 05, SAA 01, SAA 06]. The purpose of MCDA is to rationalize the choices implemented in applications of any type, as long as the relevant information linked to these choices is available, defined using multiple criteria and open to objective-based interpretation. The area of application of this technique is much broader than the field of product and system design. However, targeted studies have succeeded in isolating and treating certain characteristics, which are specific to design engineering. These specific points concern the formulation of interpretation and aggregation functions for design criteria. Important work has been carried out by Achille Messac [MES 96, CHE 00] and Antonsson [ANT 05], who generalized the formal mathematical frameworks for interpretation and aggregation. Also, it is noteworthy that in many domains of application, interpretation and aggregation are carried out *a minima*, i.e. using maximization or minimization objectives for each criterion and a global Pareto-dominance criterion. This dominance criterion is essentially mathematical in nature, and it consists of selecting all design solutions that are preferable to other solutions with regard to at least one criterion. Although this minimalist approach is widespread, it is open to criticism due to the fact that it does not take into account the designer preferences and produces a very large number of solutions, which the designer is required to sort out as part of a later selection process [GER 95].

Figure 3.4 illustrates the different steps involved in the process. The design and observation variables in the upper part of the diagram result from physical modeling. Interpretation and aggregation create a metric, which may be used to compare admissible solutions, and a solution search procedure is generally implemented using an optimization or constraint satisfaction method [BEN 06]. This approach is therefore based on the principle of equivalence between solution seeking in design and the optimization or satisfaction of a set of relationships under admissible precision constraints. Thus, for example, a design problem defined by the aggregation of desirability functions (which must be maximized) may be expressed as a search for an argument $\{X\}$ characterizing a design solution that maximizes the global preference (objective satisfaction) of the solution:

$$\arg\max X_{\text{opt}}\ (\xi \circ \delta \circ \mathbf{M}(\{X\})) \Longrightarrow \text{GDI}_{\text{perfo}}(\{X\}) \qquad [3.1]$$

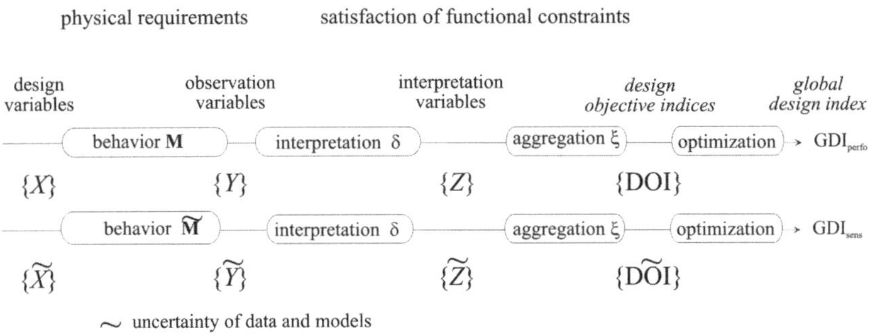

Figure 3.4. *From the observation of physical behavior to preference aggregation and the selection of the solution with the best index*

The solution of [3.1] gives the optimal vector, which satisfies the relevant objectives and constraints, but the robustness of the solution depends on the sensitivity of this optimum. A second global design index, GDI_{sens}, is constructed, integrating the uncertainty resulting from data and models throughout the chain (shown in the lower part of Figure 3.4). It would be possible to observe the performance dispersion using reliability analysis and a probability measure, but this would require unavailable information and would not be particularly useful in embodiment design terms. To avoid the requirement of this analysis, two measures are proposed: first, a variation range of $\alpha = |\max \tilde{y}\ - \min \tilde{y}|$ and, second, a tolerance with a nominal value of $\beta = |\text{mean}(\tilde{y})\ - y|$. The GDI_{sens} index combines the satisfaction of these two requirements using a desirability function: α represents the robustness aim (performance stability) and β represents the reliability objective (weak performance deviation). A design is robust if it satisfies the $\text{GDI}_{\text{perfo}}$ performance requirement and the GDI_{sens} sensitivity requirement.

The interpretation δ and aggregation ξ models combine the objectives and constraints of the problem, and have the potential to successfully define a single optimization objective. This approach considers design constraints and objectives as elements of the same nature, resulting from a requirement and a global functional model. The numerical processing of equation [3.1] poses certain difficulties linked to the solution of problems with a relatively high number of variables (number of components of vector $\{X\}$), with pairings leading to complex behavior (which cannot be calculated in polynomial time). In this domain, developments in soft computing [SAR 08] have produced efficient algorithms, which are able to take account of notions such as robustness and imprecision.

3.1.3.2. *The performance-sensitivity compromise*

The performance-sensitivity compromise is another example of an aggregation problem. Several modeling-based approaches have been proposed [SCO 98, QUI 12, QUI 13] using the notion of the general weighted average:

$$p_{GPI} = \left(\frac{w_1\, p_1^s + w_2\, p_2^s}{w_1 + w_2} \right)^{\frac{1}{s}}.$$

This requires us to determine the relationships between weights w_i and power s for preferences $p_1(\mathrm{GDI}_{\mathrm{perfo}})$ and $p_2(\mathrm{GDI}_{\mathrm{sens}})$. The compromise must be robust, i.e. present weak sensitivity to small variations in p_1 and p_2; this depends on the value of the power s.

Negative values of s guarantee all of the axioms needed in order to use the optimization function in design. More specifically, if s is negative, then the optimization function is idempotent: when a solution is evaluated (calculation of Z_G) and all of the components of vector $\{Z\}$ are worth $1/2$ (average satisfaction), then its overall value is $Z_G = 1/2$ (average satisfaction). Moreover, the optimization function satisfies an annihilation axiom: when any component of vector $\{Z\}$ is worth 0 (zero satisfaction), then the global value is $Z_G = 0$ (unacceptable solution).

s is chosen within the domain $[-\infty, 0]$ using the stability of the optimal solution calculated with a variable s (decision robustness). A calculated optimal solution may, in fact, be highly sensitive to weighting coefficients w_i, i.e. to the criticality of the different criteria used in its evaluation. A robust solution must be robust in terms of its behavior, its physical characteristics and from a decision-making perspective. [QUI 12] gives the example of the design of a flash evaporator, which is used to cool and concentrate grape juice in the wine-making process. The characteristics of the apparatus are optimized by calculation. Optimization is carried out from a combined physical and decision-making perspective. The proposed solutions offer designs for flash evaporators that perform well while offering weak sensitivity to uncertainty in relation to the physical environment and to the parameters of the global optimization function.

3.1.4. *Conclusion*

Following on from the progress made in multi-criteria analysis for mechanical system design in the first decade of the 21st Century, work is now required to better understand and model the criteria used by designers at the highest levels of reasoning. The notion of risk associated with uncertainty in design reasoning and its consequences for system performance should be taken into account at the earliest possible stage in the design process, i.e. at a stage where the dimensional characteristics and choices of materials, regulation and components are not yet fixed. Psychological aspects and the understanding of the technico-economical environment of the system are the key when considering this level of reasoning.

The early stages of the design process involve choices that must always constitute a compromise, driven by the priority levels (criticality) of each objective. The approach presented in this chapter aims to satisfy the criteria relating to both performance and sensitivity to uncertainty (robustness and reliability). However, choices made at this stage have major implications for the product, its behavior and its technico-economical viability. The subjective aspects (relating to the designer and decision makers) of the design problem are as crucial as the objective aspects (concerning the product) in determining the final results. Optimization to produce design solutions is of no use if these solutions are in conflict with the designer's own preferences.

Collignan's doctoral thesis [COL 12a, COL 12b] and [SEB 12] give examples of optimized design of systems constructed by seeking compromises between subjective aspects (preferences expressed in terms of trust levels) and objective aspects (performance). A generalization of these approaches has yet to be developed. A generalized decision support approach, combining performance, robustness and reliability indicators with more subjective criteria (limited rationality of decision makers), is one of the perspectives offered by multi-criteria decision analysis.

The engineers at Electricité de France (EDF), confronted with decision problems involving a range of issues (availability, security and safety, maintenance costs, brand image, investments, etc.), have developed a designer preference model using a utility function. This model has proved successful in identifying acceptable compromises, and it has been put to practical use in designing hydroelectric dams, nuclear power plants and wind farms [BEA 99, BEA 09] □.

3.2. Summary and conclusion

As we come to the end of our exploration of the realms of uncertainty, we will now attempt to summarize our findings and draw a number of conclusions relating to views of uncertainty from the perspectives of the user, the engineer and the researcher: three groups concerned by decision-making in the presence of uncertainty.

3.2.1. *Three perspectives*

3.2.1.1. *The user perspective*

Everyone would like to live in an impossible world of zero risk, and individuals rely on society to offer protection. In response to this requirement, public authorities issue universally applicable laws aiming to offer some guarantee of security. Individuals also, as consumers, wish to have access to robust and reliable products. These considerations may be illustrated using a principle and two French legal texts.

3.2.1.1.1. The precautionary principle

This principle, which is often affirmed without the introduction of necessary adaptations for specific situations, has been the subject of considerable reflection. Discussions in the 1990s, particularly in the wake of food and public health crises, led the European Commission to issue guidelines [COR 01] on environmental policy, which were then extended to other domains. The precautionary principle involves the perception or identification of a potential risk, for which a scientific study has given uncertain results. Politicians are then free to act, or otherwise, considering proportionality, nondiscrimination and coherency with other similar measures, and by re-examining the scientific situation.

In France, a report presented to the Prime Minister by Kourilsky and Viney [KOU 00] provided a highly detailed analysis of these ideas, concluding that it would be more helpful to think in terms of an attitude than a principle: the authors recommend replacing the idea of "if in doubt, refrain from acting" with "if in doubt, do everything possible to act in a better way".

A wide range of contributions were made over the following years, from philosophical, judicial, political and technical perspectives. Whether or not this principle is beneficial for society and its members is still open to debate. The precautionary principle constitutes a social response to the acceptability constraints of a decision.

The opening speech of the 2008 IMdR conference in Lille, France, on the theme of *risk and performance*, was given by Jean-Louis Ricaud, Senior Vice President of the Renault Group [RIC 08]. Ricaud highlighted the following two essential points concerning the precautionary principle as it relates to innovation.

– *To create is to take risks. Innovation is, by nature, a step into the unknown, and therefore a risk taken in the face of unpredictability.*

– *The principle of development is a fundamental aspect of European regulations. This principle means that we cannot be held responsible for risks that could not have been anticipated at the time of product design. Consequently, we have an obligation of means that consists, at the time of design, of ensuring that all applicable regulations have been taken into account and that, using knowledge currently available, all predictable risks have been identified, evaluated and treated.* This implies that we must use not only the entirety of our own knowledge, but all of the knowledge currently available to us.

The precautionary principle may be seen as a reformulation of a statement made by Blaise Pascal in 1654, introducing cost-to-benefit reasoning: *when we work for tomorrow, and do so on an uncertainty, we act reasonably, for we ought to work for an uncertainty according to the doctrine of chance that has been demonstrated.* We work for uncertainty because this is inherent to all human action; we work for tomorrow, as our role is to construct our world. We act reasonably in doing this, as the expected benefit for society or for individuals is greater than the cost of failure. Life, and all of our actions, takes place in a context of uncertainty, and we must live with this fact; only life itself has a certain outcome.

3.2.1.1.2. Constitutional act/Constitutional law concerning the environmental charter

Constitutional law is the highest level of legal authority in France. The text adopted on February 28, 2005 by Parliament, meeting in Congress, and promulgated on March 1, 2005 by the President of the Republic includes the following article:

Article 5: when the occurrence of damage, albeit uncertain given the state of scientific knowledge, might have a serious and irreversible effect on the environment, public authorities will ensure, by application of the precautionary principle in their relevant domains, the implementation of risk assessment procedures and the adoption of provisional and proportional measures to prevent the occurrence of this damage.

This article highlights both the risk evaluation and the definition of crisis management measures, where necessary. It reflects attitudes to decision-making in Extremistan: whatever the possibility of an event occurring, the consequences must be acceptable.

3.2.1.1.3. Prevention of technological and natural risks

The French law of July 30, 2003 concerning risk prevention, established in the wake of the Toulouse accident of September 21, 2001 and published before the constitutional text set out above, states that: *measures provided for in plans for preventing technological risks [...] are to be implemented progressively, notably in relation to the probability, gravity and kinetics of potential accidents, and to the relationship between the cost of the intended measures and the expected security benefits.*

When considering mechanics in a context of uncertainty, we have seen that probability is not always a measure of uncertainty. Nevertheless, the use of the word "probability" in a legal text constitutes the recognition of a possible technological hazard. The understanding of this term by judges in relation to mechanical scenarios has yet to be seen. The severity and kinetics of potential accidents are linked to behavior models.

A considerable number of other texts issued by public authorities lay out obligations in terms of rules, regulations and directives, and in norms codifying industrial practice and technical abilities. The principle and texts discussed above are fundamental, and demonstrate attempts to place uncertainty mastery into a legal framework, providing a collective response to the reality of uncertainty (section 1.1). Therefore, they must be considered to be essential requirements in mechanical design and decision acceptability considerations, working to avoid underprepared excursions into Extremistan. The French decree of February 7, 2012, codifying general rules relating to basic nuclear installations, demonstrates a good application of these ideas.

3.2.1.1.4. The consumer

As users, consumers, whether individuals or project managers, wish to have access to robust and reliable products, and do not tend to distinguish between the two adjectives. A thorough definition of the significance of *robust* is given in section 1.2. In addition to ensuring that products are not dangerous, in accordance with industry norms, the mastery of uncertainty in mechanics requires us to satisfy four requirements, while not exceeding cost thresholds. The first requirement is user satisfaction: the product must successfully provide the service expected by the user. Second, there must be a capacity for maintenance: it should be possible to maintain the product.

Third, regarding sustainable development, the product must be economical in terms of natural resources and energy, and must be recyclable. The fourth requirement is for guarantees: the designer makes a commitment in relation to uncertain events that may lead to failure or reduced operations.

3.2.1.2. *The engineering perspective*

In a social, economical and technical context, engineers must design products or systems that exist in a context of uncertainty from the very start of their lifecycle through design, production and use to final recycling. Designing a robust and reliable product, as discussed in 1.2, according to a design for robust production, requires our knowledge to evolve throughout this lifecycle.

3.2.1.2.1. The design cycle

Section 1.3 highlighted the need to create a design cycle in which all phases and steps are considered in the context of uncertainty, using increasingly detailed knowledge at increasingly high granularity levels for functional validation and technical verification. It is interesting to reconsider Figure 1.15, from the conclusion of section 1.3.6 (Figure 3.5). First, the design cycle requires the ability to integrate all phases and steps, from those based on the cognitive sciences to those using different branches of physics, mechanics and mathematical modelings. It is pointless to consider single aspects in isolation. For this reason, *Designing for robust production*, as explained section 1.3, is based on the use of multidisciplinary teams, developing different abilities to the same level of maturity in each phase or step.

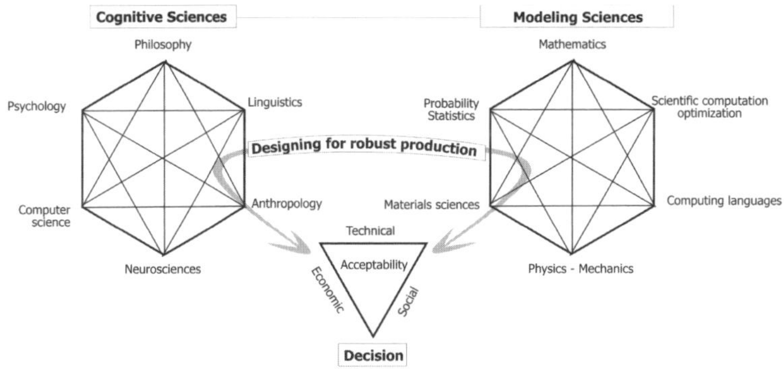

Figure 3.5. *Symbolic representation of the cognitive sciences (left) and the modeling sciences (right). Designing for robust production requires sustained and effective dialog between the two domains in order to produce acceptable decisions*

3.2.1.2.2. Maturity of knowledge

The development of knowledge concerning a product may be illustrated using the (maturity, uncertainty reduction and decision) triplet. During the design phase and up until production, successive decisions and the detailing of physical models reduce the field of possibilities. In the active use phase, development is guided by feedback.

3.2.1.2.3. Data representation

A piece of data, in this case taking the form of a variable but which may extend to a function of space and time, may be associated with a model-based representation of knowledge. This notion, introduced in Figure 2.5, can be extended to take account of possible enhancements (Table 3.1). A variable is then noted with its associated item, for example X_{item}.

Item	Representation	Parameters
1	Qualitative, descriptive	Text
2	Social acceptation	Degree
3	Environmental acceptation	Degree
4	Generalized cost	€ or $
5	Boolean affirmation	0 or 1
6	Expert value	$x_e \in \mathbb{R}$
7	Characteristic value	$x_k \in \mathbb{R}$
8	Robust interval	$\mathbf{x} \in [x_{\min}, x_{\max}]$
9	Dimension on a plan	$x \in \mathbb{R}$
10	Fabrication tolerance	Capability
11	Possibilistic	$\prod(X)$, $\mathrm{N}(X)$
12	Probabilistic	Distribution of X
13	Probability or index	P_f or $\beta \in \mathbb{R}$
14	Experimental measure	$\tilde{x} \in [x_{\min}, x_{\max}]$
⋮	...	

Table 3.1. *Generalized (generalizable) knowledge entity for a variable*

This representation is associated with the characterization level defined in section 2.2.1.2: durability (d), variation (v), sensitivity (s) and completeness (c). The designer's task is to construct the variable $X_{item}^{(d,v,s,c)}$ for each of the parameters in question.

3.2.1.2.4. Knowledge propagation

The propagation of knowledge (or uncertainty) concerns the design entity resulting from the combination of basic variables in a behavior model:

$$Y^{(d,v,s,c)} = \mathbf{M}\left(X_{item}^{(d,v,s,c)} \Big|_i \; i = 1, ..., n \right)$$

This model itself is associated with a characterization level. If the knowledge entity is homogeneous, all parameters are represented by the same item at the same level of characterization; the richness of sensitivity or reliability analysis methods was shown in section 2.3. This section also highlighted the paucity of resources when dealing with heterogeneous knowledge entities.

Particular care is needed concerning the sensitivity of the model $\mathbf{M}$ in showing weak data variations. From a mathematical perspective, sensitivity is measured by the finite difference $\frac{\Delta Y}{\Delta X}$ that tends toward $\frac{0}{0}$, an indeterminate form that is unacceptable for engineering or numerical analysis purposes. The limitations of finite element modeling quickly become clear when weak disturbances produce considerable effects. In these cases, the mesh is unable to cope with these changes and a new, specific formulation is required. Examples of this are given by a study of the stability of shells in relation to geometric defects [JUL 99] and by work on springback in die forging [LEB 13a].

3.2.1.2.5. Knowledge aggregation

Data models, behavior models and knowledge propagation involve the superposition of knowledge, parts of which may prove contradictory. Section 3.1 offered a framework for multiple-criteria optimization, used to support decisions, which always constitutes a compromise as shown in Figure 3.2 (reproduced here), as Figure 3.6. A global preference function is used to guarantee the compromise between mass and compactness requirements.

3.2.1.2.6. The TRL scale

[2]Industrial players are faced with the consequences of incomplete knowledge on a daily basis: the limitation of designs based on prior experience, the multitude of scientific and technical domains and the difficulty of mastering the complexity involved in key areas. Uncertainty may be seen as the main cause of industrial problems.

2 From a contribution by J.-M. THÉRET.

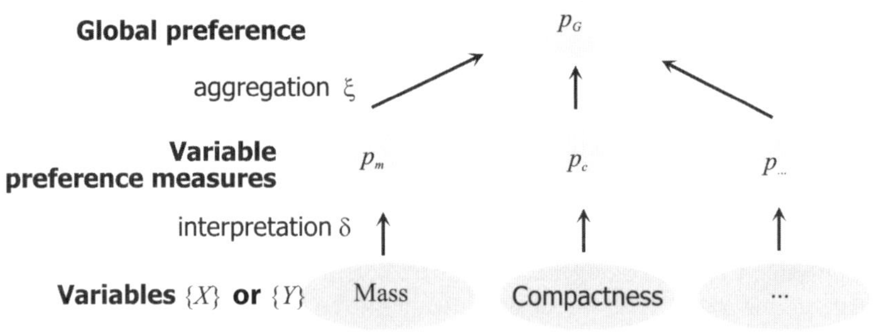

Figure 3.6. *Interpretation and aggregation of variables in a preference model*

Modeling is now widely used in the context of industrial design, with the development of static, fatigue, dynamic, thermal or fluid models used to reproduce functions, constraints and performance in a range of different forms. Models have improved due to increasing computing power, enabling the use of more complex physical laws, and the identification of better behavior laws giving a better rendering of complex real phenomena in all mechanical disciplines (fluid mechanics, thermal engineering, solid mechanics), alone and in combination. Modeling has also undergone development in the context of production procedures (forging, deep drawing, welding etc.), as the final aim of industrial activity is, clearly, the production of designed pieces.

Well-developed technological innovations always follow a clearly established, progressive workplan. The purpose of this plan is to detail the concept, demonstrate the capacity for development of the new product and acquire the knowledge required for risk-free use of the product. Regular evaluation of this technological maturity is important, and can be implemented using the technology readiness level (TRL) grid established by NASA (Figure 3.7). The grid shows the evolution of maturity in line with knowledge acquisition. The workplan adapts to this progression in a logical manner.

Maturity monitoring is directly linked to uncertainty: as technological innovation progresses, the random aspect of knowledge is reduced. Research projects aim to remove uncertainty in all areas, excepting factors, parameters or hypotheses that will not have a major effect on the lifecycle of the product in question.

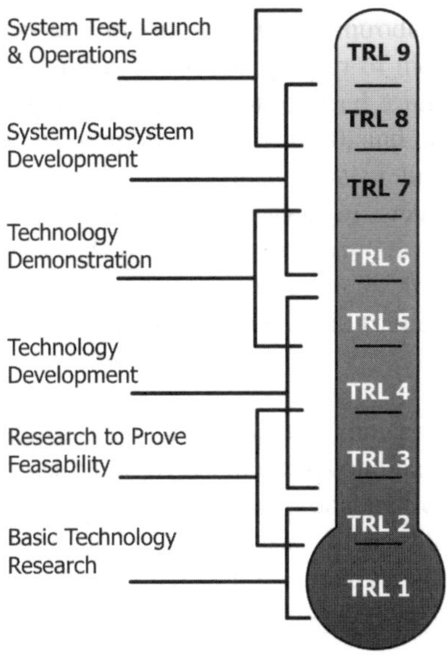

Figure 3.7. *TRL – technology readiness level scale (source: National Aeronautics and Space Administration (NASA))*

The multiplicity of industrial requirements (relating to culture, the domains in question and the interface exchanges involved) means that we must be open to a variety of solution methods. A wide range of responses have been covered in this book, in terms of both tools and scientific methodologies. Further work is now needed to develop and use these tools and methodologies in the context of industrial problems □.

3.2.1.3. *The research perspective*

The viewpoints of researchers have been well represented in this book as a result of the professional backgrounds of the authors. Two aspects will be considered in this section: top-down and bottom-up guidance.

3.2.1.3.1. Top-down guidance

Top-down guidance includes work in the fields of mechanics, physics, materials science and mathematics, domains which are well represented within the mechanics community, but also from the perspective of the cognitive sciences. Work has been carried out in relative isolation within each separate domain, often independent of design objectives. Thus, a detailed mathematical study of stochastic processes will have no impact on the mastery of uncertainty in mechanics if it is not based on reality, i.e. using available information; nevertheless, this work is essential, and should be taken into consideration in postproduction research. From a design perspective, top-down assistance consists of a simple question: *we have a solution, do you have a problem to match?*

This idea was eloquently expressed by Leonardo da Vinci: *no human investigation can be called real science if it cannot be demonstrated mathematically.*

3.2.1.3.2. Bottom-up guidance

Bottom-up guidance consists of work on issues raised by the solutions to technical problems. These projects may be focused on only one specific aspect, without sufficient consideration for the general context and without using a sufficiently robust modeling framework. They are helpful in identifying stumbling blocks for innovation, which must then be solved using a scientific approach.

Once again, we will conclude in the words of Leonardo da Vinci: *he who loves practice without theory is like the sailor who boards ship without a rudder and compass and never knows where he may cast.*

These quotations, which are displayed in the Château du Clos-Lucé museum in Amboise, France, illustrate the need for researchers to take account of contrasting perspectives. A number of priorities for researchers are suggested in section 3.2.2.

3.2.1.3.3. Creating a science of uncertainty: two pathways

The benefits of using research tools are evident when considering uncertainty from an engineering perspective. Recent developments have tackled the general problem of uncertainty quantification, with varying degrees of efficiency, using a bottom-up type approach, slowly removing restrictive hypotheses while maintaining acceptable levels of calculability. The approach discussed in section 2.3.5 is of the top-down variety, creating significant conceptual difficulties – involving theorems that require demonstration – and calculation problems. As the authors [OWH 13] highlighted in their conclusion, the creation of a science of uncertainty involves the same ambitions, which motivated A.-N. Kolmogorov in formalizing probability theory. This is an essential issue for researchers, at the meeting point of optimization theory, probability theory, computer science, statistics and, finally, mechanics in order to ensure physical coherency. The usual approach consists of aggregating the uncertainties $\{X\}$ and $\{P\}$ resulting from elementary variables relating to design and models. The accumulation of these variables leads to issues with dimensionality. A recent alternative consists of identifying uncertainty in operators. This is the case, for example, of matrices in the finite element method, as a function of the material and geometric uncertainties involved in each term. The nonparametric method proposed by Soize [SOI 00, SOI 05] seeks to identify the stochastic properties of operators satisfying the maximum entropy principle, and introduces a number of scattering parameters, identified using experimental testing.

3.2.1.4. *Innovation*

The principles involved in mastering uncertainty may appear to limit innovation; in reality, they act as safeguards for sound strategy. Innovation by evolution consists of moving slowly within the boundaries of Mediocristan, and presents minimal risks, according to the percentage of the product or system that is modified. Breakaway innovations constitute excursions into Extremestan, and in such cases it is essential to demonstrate mastery of the events associated with failure.

3.2.2. *Challenges in engineering science*

When designing for robust, reliable and durable production, we must consider a range of different topics, and the design process itself constitutes an *engineering discipline* in its own right. As with all sciences, the researcher is not only an observer, but also, in this discipline, a creator and an inventor. The aim of engineering is to participate in the creation of objects and systems and to analyze their operations. Objects and systems are made up of complex structures with a wide variety of content. Research in this area cannot be limited to the simple application of a given disciplinary field or trade, and possesses its own methodological and scientific content; it cannot be limited to the implementation of a specific engineering science, but constitutes a distinct research approach within the field of engineering. This approach involves challenges linked to understanding, modeling and decision-making. The analysis given in this section is the result of discussions that took place within the AFM (scientific and technical workgroup on "mechanics and uncertainty") and the IMdR (technical workgroup on "structural safety and security"), identifying five significant challenges.

3.2.2.1. *Understanding and modeling uncertainty*

3.2.2.1.1. Uncertain knowledge of behavior models and their parameters

First, we should note the importance of understanding, observing and obtaining high-quality measures of physical phenomena. These considerations are of fundamental importance in the investigation of behavior models and their parameters, a main focus within the mechanics community; however, insufficient consideration is given to domains of validity and bias, and overlearning, which often occurs, is incompatible with generalization. Groups and individuals working with mechanics and uncertainty must make use of the best predictive models available, on the condition that these models are associated with a domain of definition and that a measure of uncertainty is available for their parameters. It is essential to take account of the model gap (often, erroneously, referred to as the model error). The operational model is produced by the approximate numerical solution to the mathematical formulation of physical considerations, which are themselves approximated, perceived and recalibrated in relation to uncertain measures. How can we represent this gap across the domain of interest of model definition, taking account of the accumulation of different gaps, from the inaccessible physical reality to the virtual representation of this reality, using parameters, or even solutions, which are affected by uncertainty?

> *Challenge 1*: Give full consideration to uncertainty relating to the parameters of models of physical realities, concentrating on the need to associate models with tests, and propose a common and unified vocabulary for those involved in testing and numerical development to promote mutual growth, in terms of experimental simulations for the first group and in virtual testing for the second group.

3.2.2.1.2. Modeling uncertain parameters

The mechanics of uncertainty is based on data or parameters. The first issue relates to the best treatment of these data using mathematical representations. Information may be obtained by two forms of enhancement: the first uses increasingly precise and numerous measures and the second concerns the decision process. Models may include the use of expert evaluations, fuzzy numbers, possibility and necessity, evidence measures, probability boxes and random variables: random fields in space and stochastic processes in time. The connection between the information and the model is made using an evolving knowledge entity. Probabilistic models represent one of many levels of knowledge, and are based on a statistical interpretation of data. This is a powerful tool; those working in the field of mechanics need to provide statisticians with a model that takes account of the quantity of data, which is often low, and physical admissibility constraints.

Challenge 2: Explore all possible modes of representing uncertainty (from expert evaluations to probability) in relation with a given objective, focusing on the representation of dependence between data values in spatial and temporal terms without introducing additional information. Representation modes must also be associated with confidence levels: a numerical value should never be given without analyzing its significance (calculation of "error", confidence interval, etc.).

3.2.2.2. *Propagation modeling and decision-making*

3.2.2.2.1. Uncertainty propagation in models and decision-making

The knowledge model used for variables and parameters is crucial, and requires strong cooperation between information specialists (including statisticians) and those working in mechanics, based on the subjacent decision objectives. This model is connected with physical behavior models and their algorithmic representations. The pairing of knowledge and behavior constitutes the input for the two main propagation models, sensitivity analysis and reliability analysis, which are used to calculate variables of interest (reference values) resulting from design and behavior variables. Sensitivity analysis is concerned with the uncertain character of variables of interest, which is of great importance during robust design optimization. Reliability analysis considers the scenarios leading to unacceptable values for variables of interest. These two forms of analysis are interconnected, but specific techniques are used for economical implementation of each method. In recent years, mathematical studies of probability and classification techniques have resulted in considerable methodological progress using probabilized knowledge: polynomial chaos, support vector machines, kriging classification, approximation and meta-modeling methods, and non-parametric approaches. While the principles behind these methods have been established, their numerical implementation remains cumbersome outside of the realm of academic examples, when dealing with industrial applications of large stochastic dimensions. Despite remarkable progress in probabilistic propagation, it is not always possible, or even helpful, to use these tools, depending on the nature of the decision to be made, and the conditions of acceptability must be the subject of multidisciplinary reflection (e.g. from engineering, economic, sociological, philosophical and legal perspectives).

Challenge 3: Construct propagation algorithms in an uncertainty representation space, probabilized or otherwise, taking account of the dimension and the complexity required for applications. Learn to analyze the results obtained in relation to the decision being made.

3.2.2.2.2. Global performance approaches

Technological products are complex systems involving elements observed at different scales (micro, meso and macro) with multiple interactions. The propagation of uncertainty from one scale level to another, from one component to another or from one technology to another is an essential consideration for operational safety and in systems engineering in relation to the anticipation of risks and the demonstration of desired performances throughout the product lifecycle. These performances relate to availability, maintainability, reliability and safety, i.e. the capability of a system to not cause damage to goods, persons or, increasingly, the environment. Functional performance aspects (such as fuel consumption, cost, speed, etc.) must also be taken into consideration. These aspects can be considered using simulation (multiphysics, systems, etc.). A bipartite response to these considerations is needed, based on performance demonstrations and risk prevention plans. The global performance, security, safety approach is not yet widely used in mechanics. Those involved in the field need to become more aware of global performance approaches when dealing with complex systems, without necessarily requiring a detailed knowledge of cyndinics (the science of danger), which relates to other disciplinary fields.

Challenge 4: Construct a multicomponent, multiscale and multiphysical representation of uncertainty propagation, including all dimensions of the design chain, the production and the product lifecycle.

3.2.2.2.3. Combating cultural inertia

The final major challenge is of a different type and concerns the teaching of uncertainty at secondary school level, where almost all disciplines only deal with certainties, at university level, and in schools of engineering. A response to this challenge was given by Max Planck: *a new scientific truth does not triumph by convincing its opponents and making them see the light, but rather because its opponents eventually die, and a new generation grows up that is familiar with it.* The texts currently being discussed for the new program for preparatory engineering classes in France (2 years, from age 18) raise the possibility of teaching probabilities in finite spaces, in a more reality-based and less dogmatic approach, but with a focus on calculation, rendering the subject more accessible. Nevertheless, probability is only part of the issue; the concepts of mechanical systems and functional safety, i.e. technology, need to be placed on the same level as concepts of mechanical behavior.

> *Challenge 5*: Reconsider the way in which probability is taught, and introduce the teaching of operational safety in systems, starting from the object and its associated issues and working toward mathematical considerations, and not vice versa; working in the opposite direction, the object is hidden from view.

3.2.3. *Industrial issues*

Industrial issues concern all trades in which products (resources) are designed to respond to a need; often, the resource takes the form of a product with certain tolerances, and the need is expressed in the form of a specification, dictated by the user or by nature. We will now consider a number of examples:

– Around 80% of mechanical failures result from a fatigue phenomenon, originating from a surface state with random faults.

– The stability of tall structures is highly dependent on geometric defects.

– Natural events (earthquakes, swells, wind, etc.) are essentially random and the models used to describe them are often rather basic. Moreover, the physics of these events is subject to evolution as a result of global warming.

– Materials such as soil are subject to significant variations in terms of their properties.

– Even in the case of more complex materials, the resistance of parts depends on the random organization of crystalline phases and the existence of connections between these phases.

– Production tolerances are simply requirements that must be satisfied in order to meet the quality objectives of a system.

– Calculation rules define partial coefficients calibrated on probabilistic methods.

– The margins introduced by the application of these rules can lead to excessively or insufficiently conservative results.

– Maintenance is based on an analysis of situations and the prediction of damage levels. The use of models for treating uncertainty allows us to plan for adapted, and not systematic, maintenance.

All industries involving mechanical aspects are concerned by these issues: process industries, mechanical construction (automotive, aeronautic, naval and rail), civil construction, etc., either directly or via procedures resulting from uncertainty analysis, in computing codes or in databases.

Mastering uncertainty presents major advantages in determining residual margins throughout the product lifecycles and in innovation. While slow evolutions within Mediocristan make use of small extrapolations of known conditions, the situation is different for breakaway strategies that take us into Extremistan; in these cases, detailed risk analysis is essential to allow designers to take the leap.

Uncertainty is now accepted as a fact of life and an inherent part of human constructions. Scientific studies have provided a large amount of information on the subject; however, these efforts are too specific to allow the creation of a "science of design", which would be broad enough to cover all of the aspects involved in industrial product and system innovation.

The author, along with all those involved in writing this book, hopes to have contributed to encouraging cooperation between researchers and engineers from different disciplines in constructing a true science of uncertainty. We hope that this cooperation, by adding to and combining the avenues explored, will inspire new perspectives and provide answers to essential questions.

Bibliography

[ABE 10] ABE S. *Support Vector Machines for Pattern Classification – Two-Class Support Vector Machines*, Advances in Pattern Recognition, Springer, London, 2010.

[ACH 10] ACHIBI M., BRONIATOWKI M., "Modèle de cox bivariés et copules", *17e Congrès de Maîtrise des Risques et de Sûreté de Fonctionnement, La Rochelle,* 5–7 October, 2010.

[ACH 12] ACHIBI M., Estimation des probabilités associées aux événements rares en aéronautique, PhD Thesis, Pierre and Marie Curie University, 2012.

[AFI 13] AFIS., Available at: http://www.afis.fr/default.aspx.

[AGU 12] AGUIRRE-MARTINEZ F., Reliability analysis of systems using belief functions theory to represent epistemic uncertainty, PhD Thesis, University of Technology of Compiègne, 2012.

[AGU 13] AGUIRRE F., SALLAK M., SCHÖN W., "Incertitudes aléatoires et épistémiques, comment les distinguer et les manipuler dans les études de fiabilité?", *10ème Congrès International Pluridisciplinaire en Qualité et Sûreté de Fonctionnement: QUALITA 2013*, 2013.

[AMM 06] AMMAR A., MOKDAD B., CHINESTA F., *et al.*, "A new family of solvers for some classes of multidimensional partial differential equations encountered in kinetic theory modeling of complex fluids", *Journal of Non-Newtonian Fluid Mechanics*, vol. 139, pp. 153–176, 2006.

[ANT 05] ANTONSSON E.K., SEBASTIAN H.J., "Fuzzy fitness functions applied to engineering design problems", *European Journal of Operational Research*, vol. 166, pp. 794–811, 2005.

[ASK 85] ASKEY R., WILSON J., "Some basic hypergeometric polynomials that generalize jacobi polynomials", *Memoirs American Mathematical Society*, vol. 319, pp. 1–50, 1985.

[AU 01] AU S.K., BECK J.L., "Estimation of small failure probabilities in high dimensions by subset simulation", *Probabilistic Engineering Mechanics*, vol. 16, 2001.

[BAB 04] BABUSKA I., TEMPONE R., ZOURARIS G.E., "Galerkin finite element approximation of stochastic elliptic partial differential equations", *SIAM Journal of Scientific Computing*, vol. 24, no. 2, pp. 619–644, 2004.

[BAR 06] BAROTH J., BODÉ L., BRESSOLETTE PH., *et al.*, "SFE method using hermite polynomials: an approach for solving nonlinear mechanical problems with uncertain parameters", *Computer Methods in Applied Mechanics and Engineering*, vol. 195, nos. 44–47, pp. 6479–6501, 2006.

[BAS 10] BASUDHAR A., MISSOUM S., "An improved adaptive sampling scheme for the construction of explicit boundaries", *Structural and Multidisciplinary Optimization*, vol. 42, pp. 517–529, 2010.

[BAU 06a] BAUDRIT C., DUBOIS D., "Practical representations of incomplete probabilistic knowledge", *Computational Statistics & Data Analysis*, vol. 51, pp. 86–108, 2006.

[BAU 06b] BAUDRIT C., DUBOIS D., GUYONNET D., "Joint propagation and exploitation of probabilistic and possibilistic information in risk assessment", *IEEE Transactions on Fuzzy Systems*, vol. 14, no. 5, pp. 593–608, 2006.

[BAU 07] BAUDRIT C., COUSO I., DUBOIS D., "Joint propagation of probability and possibility in risk analysis: towards a formal framework", *International Journal of Approximate Reasoning*, vol. 45, pp. 82–105, 2007.

[BAY 95] BAYART D., *Des savoirs en action – Des objets qui solidifient une théorie: l'histoire du contrôle statistique de fabrication*, L'Harmattan, 1995.

[BEA 99] BEAUDOUIN F., MUNIER B., SERQUIN Y., "Multi attribute decision making and generalized expected utility in nuclear power plant maintenance", *Interactions and Preferences in Decision Making*, Kluwer Academic Publishers, pp. 341–357, 1999.

[BEA 09] BEAUDOUIN F., MUNIER B., "A revision of industrial risk management", *Risk and Decision Analysis*, vol. 1, pp. 3–20, 2009.

[BEC 12] BECT J., GINSBOURGER D., LI L., *et al.*, "Sequential design of computer experiments for the estimation of a probability of failure", *Statistics and Computing*, vol. 22, no. 3, pp. 773–793, 2012.

[BEN 70] BENJAMIN J.R., CORNELL A., *Probability, Statistics and Decision for Civil Engineers*, McGraw-Hill, 1970.

[BEN 06] BENHAMOU F., GRANVILLIERS L., "Continuous and interval constraints", *Handbook of Constraint Programming*, Elsevier, pp. 571–604, 2006.

[BER 92] BERGER J., *Statistical Decision Theory and Bayesian Analysis*, 2nd ed., Springer, New York, 1992.

[BER 98] BERCHTOLD A., *Chaînes de Markov et modèles de transition*, Hermès, 1998.

[BER 06] BERVEILLER M., SUDRET B., LEMAIRE M., "Stochastic finite element: a non intrusive approach by regression", *European Journal of Computational Mechanics*, vol. 15, no. 1–3, pp. 81–92, 2006.

[BER 12] BERVEILLER M., LE PAPE Y., SUDRET B., *et al.*, "Updating the long-term creep strains in concrete containment vessels by using mcmc simulation and polynomial chaos expansions", *Structure and Infrastructure Engineering*, vol. 8, no. 5, pp. 425–440, May 2012.

[BEY 07] BEYER H.G, SENDHOFF B., "Robust optimization – a comprehensive survey", *Computer Methods in Applied Mechanics and Engineering*, vol. 196, pp. 3190–3218, 2007.

[BIC 08] BICHON B., ELDRED M., SWILER L., *et al.*, "Efficient global reliability analysis for nonlinear implicit performance functions", *AIAA Journal*, vol. 46, pp. 2459–2468, 2008.

[BOS 92] BOSER B.E., GUYON I.M., VAPNIK V.N., "A training algorithm for optimal margin classifiers", *Fifth Annual Workshop on Computational Learning Theory*, Pittsburgh, ACM, pp. 144–152, 1992.

[BOR 10] BOREUX J.J., PARENT E., BERNIER J., *Pratique du calcul Bayésien*, Springer, 2010.

[BOU 11] BOURINET J.M., DEHEEGER F., LEMAIRE M., "Assessing small failure probabilities by combined subset simulation and support vector machines", *Structural Safety*, vol. 33, pp. 343–353, 2011.

[BOU 13] BOUISSOU M., Available at: http://marc.bouissou.free.fr/siterb/default.htm.

[BOU 10] BOUSSOUARA K., Etude des émissions polluantes et des moyens de dépollution dans les moteurs à combustion interne, PhD Thesis, University of Constantine, 2010.

[BRE 09] BRÉMAUD P., *Initiation aux Probabilités et aux Chaînes de Markov*, Springer-Verlag, Berlin, Heidelberg, 2009.

[BUR 98] BURGES C.J.C., "A tutorial on support vector machines for pattern recognition", *Data Mining and Knowledge Discovery*, vol. 2, pp. 121–167, 1998.

[CAM 47] CAMERON H., MARTIN W., "The orthogonal development of nonlinear functionals in series of Fourier-Hermite functional", *Annals of Mathematics*, vol. 48, pp. 385–392, 1947.

[CAN 13] CANU S., "Machines à noyaux pour l'apprentissage statistique", *Techniques de l'ingénieur, Traitement du signal: bases théoriques*, base documentaire: TIB295DUO, 2013.

[CAP 05] CAPPÉ O., MOULINES E., RYDÉN T., *Inference in Hidden Markov Models*, Springer Series in Statistics, Springer, 2005.

[CAR 98] CARRÈRE J.C., *Le cercle des menteurs, contes philosophiques du monde entier*, Plon, 1998.

[CEN 03] CEN. Eurocode: bases de calcul des structures, Technical Report NF EN 1990, AFNOR, March 2003.

[CHE 00] CHEN W., SAHAI A., MESSAC A., *et al.*, "Exploring the effectiveness of physical programming in robust design", *ASME Journal of Mechanical Design*, vol. 122, no. 4, pp. 385–394, 2000.

[CHE 07] CHEN K.Y., "Forecasting systems reliability based on support vector regression with genetic algorithms", *Reliability Engineering & System Safety*, vol. 92 no. 4, pp. 423–432, 2007.

[CHO 04] CHOI S., GRANDHI R., CANFIELD R., *et al.*, "Polynomial chaos expansion with Latin hypercube sampling for estimating response variability", *AIAA Journal*, vol. 45, pp. 1191–1198, 2004.

[CHO 07] CHOI S.K., GRANDHI R.V., CANFIELD R.A., *Reliability-Based Structural Design*, Springer, 2007.

[CHR 07] CHRISSIS M.B., KONRAD M., SHRUM S., *Guidelines for Process Integration and Product Improvement*, CMMI, 2nd ed., Addison Wesley, 2007.

[CLA 91] CLARK K.B., FUJIMOTO T., *Product Development Performance. Strategy, Organization and Management in the World Auto Industry*, Harvard Business Press, Boston, MA., 1991.

[CLE 00] CLÉMENT B., Conception Robuste: les concepts et la méthodologie de Taguchi dans le design de produits, Technical report, École Polytechnique de Montréal et Génistat Conseils Inc., 2000.

[COD 13] CODATA., Available at: http//en.wikipedia.org/wiki/codata.

[COL 06] COLLECTIF., Probabilistic structural integrity of PWR reactor pressure vessel – round robins on probabilistic approach for structural integrity of reactor pressure vessels, Technical report, OECD-NEA Workshop, Lyon, France, 2006.

[COL 11] COLLECTIF., Groupe de travail Management, Méthodes, Outils, Standards (M2OS) – Fiches Pratiques, Technical report, Institute of Risk Management, 25 May 2011.

[COL 12a] COLLIGNAN A., Méthode d'optimisation et d'aide à la décision en conception mécanique: application à une structure aéronautique, PhD Thesis, University of Bordeaux, I2M, UMR CNRS 5295, 2012.

[COL 12b] COLLIGNAN A., SÉBASTIAN P., PAILHES J., *et al.*, "Arc-elasticity and hierarchical exploration of the neighborhood of solutions in mechanical design", *Advanced Engineering Informatics*, vol. 26, no. 3, pp. 603–617, 2012.

[COM 93] COMBA J.L.D, STOLFI J., "Affine arithmetic and its applications to computer graphics", *Proceedings of SIBGRAPI'93 – VI Simposio Brasileiro de Computacao Grafica e Processamento de Imagens*, pp. 9–18, 1993.

[COR 01] CORCELLE G., "La perspective communautaire du principe de précaution", *Revue du Marché commun et de l'Union Européenne*, vol. 450, July–August 2001.

[COW 98] COWELL R., "Introduction to inference for bayesian networks", in JORDAN M.I. (ed.), *Learning in Graphical Models*, MIT Press, pp. 9–26, 1998.

[COX 97] COX D.D., JOHN S., "SDO: A statistical method for global optimization", In ALEXANDROV N.M., HUSSAINI M.Y. (eds), *Multidisciplinary Design Optimization: State-of-the-Art*, SIAM, Philadelphia, PA, pp. 315–329, 1997.

[CRE 09] CRESTAUX T., LE MAITRE O., MARTINEZ J.M., "Polynomial chaos expansion for sensitivity analysis", *Reliability Engineering and System Safety*, vol. 94, no. 7, pp. 619–644, 2009.

[CRI 00] CRISTIANINI N., SHAWE-TAYLOR J., *An Introduction to Support Vector Machines: And Other Kernel-Based learning Methods*, Cambridge University Press, New York, 2000.

[DAI 12] DAI H.Z., ZHANG H., WANG W., *et al.*, "Structural reliability assessment by local approximation of limit state functions using adaptive Markov chain simulation and support vector regression", *Computer-Aided Civil and Infrastructure Engineering*, vol. 27, no. 9, pp. 676–686, 2012.

[DAN 09] DANTAN J.Y., Contribution à l'optimisation de la robustesse des produits par la maîtrise des variations géométriques en phase de conception, Mémoire d'HDR, ENS Cachan, 19 June 2009.

[DAN 13] DANTAN J.Y., GAYTON N., QURESHI A.J., *et al.*, "Tolerance analysis approach based on the classification of uncertainties (aleatory/epistemic)", *Procedia CIRP*, vol. 10, pp. 287–293, 2013.

[DAR 01] DARSES F.F., DETIENNE F., VISSER W., "Assister la conception: perspectives pour la psychologie cognitive ergonomique", *Acte des journées d'étude en psychologie ergonomique*, Nantes, IRCCyN, pp. 29–30, Octobre 2001.

[DE 08] DE ROCQUIGNY E., DEVICTOR N., TARANTOLA S., *Uncertainty in Industrial Practice – A Guide to Quantitative Uncertainty Management*, John Wiley & Sons, New York, 2008.

[DEH 07] DEHEEGER F., LEMAIRE M., "Support vector machine for efficient subset simulations: ^{2}SMART method", in KANDA J., TAKADA T., FURUTA, H. (eds), *ICASP 10 – Applications of Statistics and Probability in Civil Engineering*, CERRA, Taylor & Francis, Leiden, The Netherlands, + CD-ROM. vol 1, pp. 259–260, pp. 221–228, 2007.

[DEH 08a] DEHEEGER F., BOURINET J.M., LEMAIRE M., *et al.*, "^{2}SMART: SVM et subset pour la fiabilité", *JFMS2008* , Journées Fiabilité des Matériaux et des Structures - Nantes, pp. 155–156, 26–28, March 2008.

[DEH 08b] DEHEEGER F., Couplage mécano-fiabiliste: ^{2}SMART – méthodologie d'apprentissage stochastique en fiabilité, PhD Thesis, Blaise Pascal University, 2008.

[DEM 67] DEMPSTER A.P., "Upper and lower probabilities induced by a multivalued mapping", *Annals of Mathematical Statistics*, vol. 38, pp. 325–339, 1967.

[DER 80] DERRINGER G., SUICH R., "Simultaneous optimization of several response variables", *Journal of Quality Technology*, vol. 12, pp. 214–219, 1980.

[DER 98] DERRAS C., Formalisation de l'imprécision informationnelle et des incertitudes décisionnelles des connaissances expertes pour la génération de processus de fabrication, PhD Thesis, Henri Poincaré University 11 December 1998.

[DIT 96] DITLEVSEN O., MADSEN H.O., *Structural Reliability Methods*, John Wiley & Sons, New York, 1996.

[DRO 02] DROESBEKE J.J., FINE J., SAPORTA G., *Méthodes Bayésiennes en Statistique*. TECHNIP, Paris, 2002.

[DOO 07] DOOSTAN A., GHANEM R.G., RED-HORSE J., "Stochastic model reduction for chaos representations", *Comput. Methods Appl. Mech. Engrg.*, vol. 196, pp. 3951–3966, 2007.

[DUT 09] DUTFOY A., LEBRUN R., "Practical approach to dependence modelling using copulas", *Proceedings of the Institution of Mechanical Engineers, Part O: Journal of Risk and Reliability*, vol. 223, no. 4, pp. 347–361, 2009.

[DU 04] DU L., CHOI K.K., "A new fuzzy analysis method for possibility-based design optimization", *10th AIAA/ISSMO Multidisciplinary Analysis and Optimization Conference*, Albany, NY, August. 30–September 1, 2004.

[DUB 78] DUBOIS D., PRADE H., "Operations on fuzzy numbers", *International Journal of Systems Science*, vol. 9, no. 6, pp. 613–626, 1978.

[DUB 00] DUBOIS D., NGUYEN H.T., PRADE H., "Possibility theory, probability and fuzzy sets: misunderstandings, bridges and gaps", in DUBOIS D. & PRADE H. (eds), *Fundamentals of Fuzzy Sets*, Kluwer, pp. 343–438, 2000.

[DUB 11a] DUBOURG V., SUDRET B., BOURINET J.-M., "Reliability-based design optimization using kriging and subset simulation", *Struct. Multidisc. Optim.*, vol. 44, no. 5, pp. 673–690, 2011.

[DUB 11b] DUBOURG V., Adaptive surrogate models for reliability analysis and reliability-based design optimization, PhD Thesis, Blaise Pascal University, Clermont II, 2011.

[DUB 12] DUBOURG V., DEFAUX G., WILLAUME P., *et al.*, "Échantillonnage préférentiel quasi-optimal par krigeage pour l'évaluation de la fiabilité des cuves de REP", *Proc. 7e Journées Fiabilité des Matériaux et des Structures*, Chambéry, 2012.

[DUB 13] DUBOURG V., DEHEEGER F., SUDRET B., "Metamodel-based importance sampling for structural reliability analysis", *Probabilistic Engineering Mechanics*, vol. 33, pp. 47–57, 2013.

[ECH 11] ECHARD B., GAYTON N., LEMAIRE M., "AK-MCS: an active learning reliability method combining kriging and Monte Carlo simulation", *Structural Safety*, vol. 33, pp. 145–154, 2011.

[ECH 12] ECHARD B., Evaluation par krigeage de la fiabilité des structures sollicitées en fatigue, PhD Thesis, Blaise Pascal University, 2012.

[ECH 13] ECHARD B., GAYTON N., LEMAIRE M., *et al.*, "A combined importance sampling and kriging reliability method for small probabilities with time demanding numerical models", *Reliability Engineering and System Safety*, vol. 111, pp. 232–240, 2013.

[EFR 79] EFRON B., "Bootstrap Methods: another Look at the Jackknife", *The Annals of Statistics*, vol. 7, no. 1, pp. 1–26, 1979.

[EL 91] EL-TAWIL K., MUZEAU J.P., LEMAIRE M., "Reliability method to solve mechanical problems with implicit limit state", in RACKWITZ R., THOFT-CHRISTENSEN P., (eds), *Reliability and Optimization of Structures, Proceedings. of the 4th IFIP WG 7.5 Conference*, Springer-Verlag, pp. 181–190, 1991.

[ENE 94a] ENEVOLDSEN I., FABER M.H., SØRENSEN J.D., "Adaptative response surface techniques in reliability estimation", in SCHUËLLER G.I., SHINOZUKA M., YAO J.T.P., (eds), *Structural Safety and Reliability*, pp. 1257–1264, 1994.

[ENE 94b] ENEVOLDSEN J., SORENSEN J.D., "Reliability-based optimization in structural engineering", *Structural Safety*, vol. 15, pp. 169–196, 1994.

[EPP 97] EPPINGER S.D., KRISHNAN V., WHITNEY D.E., "A model-based framework to overlap product development activities", *Management Science*, vol. 43, no. 4, pp. 437–451, 1997.

[FAR 89] FARAVELLI L., "Response surface approach for reliability analysis", *Journal of Engineering Mechanics*, vol. 115, no. 12, 1989.

[FER 03] FERSON S., GINZBURG L., KREINOVICH V., *et al.*, Construction of probability boxes and Dempster-Shafer structures, Technical Report SANDD Sandia National Laboratories, pp. 2002–4015, 2003.

[FIG 04] DE FIGUEIREDO L.H, STOLFI J., "Affine arithmetic: concepts and applications", *Numerical Algorithms*, vol. 37, pp. 147–158, 2004.

[FOR 91] FORSBERG K., MOOZ H., "The relationship of system engineering to the project cycle", *Procedings of INCOSE Symposium*, pp. 57–65, 1991.

[GAB 01] GABLONSKY J.M., em Modifications of the DIRECT algorithm, PhD Thesis, North Carolina State University, 2001.

[GAY 02] GAYTON N., BOURINET J.M., LEMAIRE M., "CQ2RS: a new statistical approach to the response surface method for reliability analysis", *Structural Safety*, vol. 25, pp. 99–121, 2002.

[GAY 09] GAYTON N., DUC E., LEMAIRE M., "The use of probabilistic approaches for the tolerancing management in design process", *Geometric Variations within Product Life-Cycle Management*, CIRP - CAT Conference, Annecy, pp. 8, 26–27, March 2009.

[GHA 91] GHANEM R.G., SPANOS P.D., *Stochastic Finite Elements: A Spectral Approach*, Springer, Berlin, 1991.

[GOH 05] GOH Y.M., BOOKER J.D., MCMAHON C.A., "A framework for the handling of uncertainty in engineering knowledge management to aid product development", *Proceedings ICED05, International Conference on Engineering Design*, Melbourne Australie, pp. 15–18, August 2005.

[GRE 05] GRECO S., "Multiple Criteria Decision Analysis: State of the Art Surveys", *International Series in Operations Research & Management Science*, vol. 78, Springer, 2005.

[GRE 06] GREBICI K., OUERTANI M.Z., BLANCO E., *et al.*, "Conflict management in design process: focus on change impact", *Actes de CE06, International Conference on Concurrent Engineering*, Antibes, France, September 2006.

[GRE 07] GREBICI K., La Maturité de l'information et le Processus de Conception Collaborative, PhD Thesis, National Polytechnic Institute of Grenoble, pp. 25–37, 2007.

[GER 95] GERO J.S., SUSHIL J.L., "Improving Pareto optimal designs using genetic algorithms", *Computer-Aided Civil and Infrastructure Engineering*, vol. 10, no. 4, pp. 239–247, 1995.

[HAN 97] HANSSEN R.W., "Reducing delivery times in engineer-to-order firms by using the concepts of concurrent engineering", *Proceedings of the 4th International Conference on Concurrent Enterprising (ICE 97)*, The University of Nottingham, pp. 495–508, October 1997.

[HAI 90] BEN-HAIM Y., ELISHAKOFF I., *Convex Models of Uncertainty in Applied Mechanics*, Elsevier, 1990.

[HAI 95] BEN-HAIM Y., "A non-probabilistic measure of reliability of linear systems based on expansion of convex models", *Structural Safety*, vol. 17, pp. 91–109, 1995.

[HAI 04] BEN-HAIM Y., HEMEZ F.M., "Robustness, fidelity and prediction-looseness of models", *Proceedings of the 7th Biennial Conference on Engineering Systems Design and Analysis*, 2004.

[HAR 65] HARRINGTON E.C., "The desirability function", *Industrial Quality Control*, vol. 21 no. 10, pp. 494–498, 1965.

[HEY 88] HEYWWOD J.B., *Internal Combustion Engine Fundamentals*, McGraw-Hill, 1988.

[HUR 04] HURTADO J.E., "An examination of methods for approximating implicit limit state functions from the view point of statistical learning theory", *Structural Safety*, vol. 26, pp. 271–293, 2004.

[INS 80] Institut de France and Académie des Sciences. *Les Sciences mécaniques et l'Avenir industriel de la France*, La Documentation française, Paris, 1980.

[IPV 06] Institut für Parallele und Verteilte Höchsleistungsrechner – IPVR – Stuttgart Universität. *Stuttgart Neural Network Simulator (SNNS)*, 2006. Available at: http://www-ra.informatik.uni-tuebingen.de/SNNS/ edition

[JEO 09] JEONG I.J., KIM K.J., "An interactive desirability function method to multiresponse optimization", *European Journal of Operational Research*, vol. 195, pp. 412–426, 2009.

[JON 93] JONES D.R., PERTTUNEN C.D., STUCKMAN B.E., "Lipschitzian optimization without the Lipschitz constant", *Journal of Optimization Theory and Applications*, vol. 79, no. 1, pp. 157–181, 1993.

[JUL 99] JULLIEN J.F., COMBESCURE A., BAILLIS C., "STANLAX, logiciel 2D d'analyse 3D de stabilité non-linéaire des coques de révolution", Actes 4ème Colloque National en Calcul des Structures – Giens, pp 889–894, 1999.

[KAR 46] KARHUNEN K., "Zur Spektraltheorie stochastischer Prozesse", *Ann. Acad. Sci. Fennicae*, vol. 37, 1946.

[KAU 91] KAUFMANN A., GUPTA M.M., *An Introduction to Fuzzy Arithmetic: Theory and Applications*, Van Nostrand Reinhold Company Inc., 1991.

[KAY 05] KAYMAZ I., "Application of kriging method to structural reliability problems", *Structural Safety*, vol. 27, no. 2, pp. 133–151, 2005.

[KEY 78] KEYSER S.J., MILLER G.A., WALKER E., Cognitive science in 1978, report submitted to the Alfred P. Sloan Foundation, New York, 1978.

[KHA 02] KHARMANDA G., MOHAMED A., LEMAIRE M., "Efficient reliability-based design optimization using hybrid space with application to finite element analysis", *Structural and Multidisciplinary Optimization*, vol. 24, no. 3, pp. 233–245, September 2002.

[KHA 08] KHAROUBI-RAKOTOMALALA C., "Les fonctions copules en finance", in *Sorbonensia Oeconomica*, Publications de la Sorbonne, 2008.

[KIM 97] KIM S.H., NA S.W., "Response surface method using vector projected sampling points", *Structural Safety*, vol. 19, no. 1, pp. 3–19, 1997.

[KIM 00] KIM K.J., LIN D.K.J., "Simultaneous optimization of mechanical properties of steel by maximizing exponential desirability functions", *Journal of the Royal Statistical Society - Series C (Applied Statistics)*, vol. 49, no. 3, pp. 311–325, 2000.

[KIU 09] DER KIUREGHIAN A., DITLEVSEN O., "Aleatory or epistemic? Does it matter?", *Structural Safety*, vol. 31, pp. 105–112, 2009.

[KLE 08] KLEIN E., *Il était sept fois la révolution*, Flammarion, 2008.

[KRE 96] KREINOVICH V., NESTEROV V.M., ZHELUDEVA N.A., "Interval methods that are guaranteed to underestimate (and the resulting new justification of Kaucher arithmetic)", *Reliable Computing*, vol. 2, pp. 119–124, 1996.

[KRY 01] KRYSL P., LALL S., MARSDEN J.E., "Dimensional model reduction in non-linear finite element dynamics of solids and structures", *International Journal for Numerical Methods in Engineering*, vol. 51, pp. 479–504, 2001.

[KRI 51] KRIGE D., "A statistical approach to some basic mine valuation problems on the Witwatersrand", *Journal of the Chemical, Metallurgical and Mining Society of South Africa*, vol. 52, no. 6, pp. 119–139, 1951.

[KRI 96] KRISHNAN V., "Managing the simultaneous execution of coupled phases in concurrent product development", *IEEE Transactions on Engineering Management*, vol. 43, no. 2, pp. 210–217, 1996.

[KOU 00] KOURILSKY P., VINEY G., *Le Principe de Précaution*, La Documentation française, 2000.

[KUS 64] KUSHNER H.J., "A new method of locating the maximum of an arbitrary multipeak curve in the presence of noise", *Journal of Basic Engineering*, vol. 86, pp. 97–106, 1964.

[LAD 08] LADEVÈZE P., "Strict upper error bounds on computed outputs of interest in computational structural mechanics", *Computational Mechanics*, vol. 42, no. 2, pp. 271–286, 2008.

[LEB 08] LEBRUN R., DUTFOY A., "An innovating analysis of the Nataf transformation from the copula viewpoint", *Probabilistic Engineering Mechanics*, 2008.

[LEB 13a] LEBON J., LE QUILLIEC G., BREITKOPF P., *et al.*, "A two-pronged approach for springback variability assessment using sparse polynomial chaos expansion and multi-level simulations", *International Journal of Material Forming*, pp. 1–13, March 2013.

[LEB 13b] LEBON J., LE QUILLIEC G., BREITKOPF P., *et al.*, "A two-pronged approach for springback variability assessment using sparse polynomial chaos expansion and multi-level simulations", *Key Engineering Materials*, vol. 554–557, pp. 957–965, March 2013.

[LEB 13c] LEBRUN R., Contributions à la modélisation de la dépendance stochastique, PhD Thesis, Université Paris VII - Denis Diderot, 24 May 2013.

[LEE 90] LEE C.C., "Fuzzy logic in control systems: fuzzy logic controller", *IEEE Transactions on Systems, Man, and Cybernetics*, vol. 20, no. 2, pp. 419–435, 1990.

[LEE 06] LEE D.U., GAFFAR A.A., CHEUNG R.C.C., *et al.*, "Constantinides Accuracy-guaranteed bit-width optimization", *IEEE Transactions on Computer-Aided Design of Integrated Circuits and Systems*, vol. 25, pp. 1990–2000, 2006.

[LEM 00] LEMAIRE M., MOHAMED A., "Finite element and reliability: a happy marriage?", in NOWAK A., SZERSZEN M., (eds), *Reliability and Optimization of Structural Systems*, Ann Arbor, *9th IFIP WG 7.5 Working Conference*, The University of Michigan, keynote lecture, pp. 3–14, September 2000.

[LEM 09] LEMAIRE M., *Structural Reliability*, ISTE, London, and John Wiley & Sons, New York, 2009.

[LEM 10] LEMAIRE M., "Concevoir Robuste et Durable: Invitation à une réflexion pour un projet pluridisciplinaire", *Association Française de Mécanique - Concevoir pour produire robuste*, French Association of Mechanics, 8 April 2010.

[LEM 12] LEMAIRE M., LANNOY A., "Perception et acceptation des risques en sécurité et sûreté des structures", *IMdR INFO (www.imdr.fr Lettre d'information trimestrielle)*, vol. 21, pp. 7–8, 2012.

[LIG 06] LIGERON J.C., *Le Cercle des Fiabilistes disparus ou Critique de la Raison Fiabiliste*, Edition préventique, 2006.

[LIN 08a] LIN S.W., LEE Z.J., CHEN S.C., *et al.*, "Parameter determination of support vector machine and feature selection using simulated annealing approach", *Applied Soft Computing*, vol. 8, no. 4, pp. 1505–1512, 2008.

[LIN 08b] LIN S.W., YING K.C., CHEN S.C., *et al.*, "Particle swarm optimization for parameter determination and feature selection of support vector machines", *Expert Systems with Applications*, vol. 35, no. 4, pp. 1817–1824, 2008.

[LOE 55] LOÈVE M.M., *Probability Theory*, Van Nostrand, Princeton, N.J., 1955.

[MAR 93] MARCOVICI C., LIGERON J.C., *Utilisation des techniques de fiabilité en mécanique*, Tec & Doc, 1993.

[MAT 73] MATHERON G., "The intrinsic random functions and their applications", *Advances in Applied Probability*, vol. 5, no. 3, pp. 439–468, 1973.

[MAT 05] MATTHIES H.G., KEESE A., "Galerkin methods for linear and nonlinear elliptic stochastic partial differential equations", *Comput. Methods Appl. Mech. Engrg*, vol. 194, pp. 1295–1331, 2005.

[MAT 07] MATHIEU L., Contribution à la maîtrise des variations géométriques des produits le long de leur cycle de vie, Habilitation à diriger des recherches, ENS Cachan – LURPA, 2007.

[MAT 11] MATTRAND C., Approche probabiliste de la tolérance aux dommages – Application au domaine aéronautique, PhD Thesis, Blaise Pascal University, 2011.

[MAT 11a] MATTRAND C., BOURINET J.-M., "Random load sequences and stochastic crack growth based on measured load data", *Engineering Fracture Mechanics*, vol. 78, no. 17, pp. 3030–3048, December 2011.

[MAT 11b] MATTRAND C., BOURINET J.M., LEMAIRE M., *et al.*, "Modeling and simulation of stochastic fatigue load sequences derived from in-flight load data", *Proceedings of the 13th AIAA Non-Deterministic Approaches Conference*, Denver, CO, 2011.

[MCD 98] MCDIARMID C., *Probabilistic Methods for Algorithmic Discrete Mathematics*, em Algorithms and Combinatorics, vol. 16, Springer-Verlag, pp. 195–248, 1998.

[MES 96] MESSAC A., "Physical programming: effective optimization for design", *AIAA Journal*, vol. 34, no. 1, p. 149, 1996.

[MER 93] MERRIAM-WEBSTER., *Merriam-Webster's Collegiate Dictionary*, 10th ed., Springfield, MA, 1993.

[MOO 79] MOORE R.E., *Methods and Applications of Interval Analysis*, SIAM, 1979.

[MOS 07] MOST T., "An adaptive response surface approach for structural reliability analyses based on support vector machines", in TOPPING B.H.V. (ed.), *Proceedings of the 11th International Conference on Civil, Structural and Environmental Engineering Computing*, St Julians, Malta, Civil-Comp Press, 2007.

[NEL 99] NELSEN R.B., *An Introduction to Copula*, Springer, 1999.

[NOT 10] NOTIN A., GAYTON N., DULONG J.L., *et al.*, "RPCM: a strategy to perform reliability analysis using polynomial chaos and resampling – application to fatigue design", *European Journal of Computational Mechanics*, vol. 19, no. 8, pp. 793–828, 2010.

[NOU 08] NOUY A., "Generalized spectral decomposition method for solving stochastic finite element equations: invariant subspace problem and dedicated algorithms", *Comput. Methods Appl. Mech. Engrg.*, vol. 197, pp. 4718–4736, 2008.

[NOU 10] NOUY A., "A priori model reduction through Proper Generalized Decomposition for solving time-dependent partial differential equations", *Comput. Methods Appl. Mech. Engrg.*, vol. 199, pp. 1603–1626, 2010.

[OTT 93] OTTO K., ANTONSSON E., "The method of imprecision compared to utility theory for design selection problems", *ASME/DTM '93: Design Theory and Methodology*, 1993.

[OWH 13] OWHADI H., SCOVEL C., SULLIVAN T.J., *et al.*, "Optimal uncertainty quantification", *SIAM Review*, vol. 55, no. 2, pp. 271–345, 2013.

[PAI 06] PAI P.F., "System reliability forecasting by support vector machines with genetic algorithms", *Mathematical and Computer Modelling*, vol. 43 no. 3–4, pp. 262–274, 2006.

[PIC 10] PICHENY V., GINSBOURGER D., ROUSTANT O., *et al.*, "Adaptive designs of experiments for accurate approximation of a target region", *Journal of Mechanical Design*, vol. 132, no. 7, 071008, 2010.

[PAI 06] PAI P.F., HONG W.C., "Software reliability forecasting by support vector machines with simulated annealing algorithms", *Journal of Systems and Software*, vol. 79, no. 6, pp. 747–755, 2006.

[PIC 75] PICKANDS J., "Statistical inference using extreme order statistics", *Annals of Statistics*, vol. 3, pp. 119–131, 1975.

[PRI 05] PRICE K.V., STORN R.M., LAMPINEN J.A., *Differential Evolution: A Practical Approach to Global Optimization*, Springer-Verlag, Berlin, 2005.

[QUR 11] QURESHI A.J., Contributions à la maîtrise de la robustesse des produits: formalisation par logique formelle, applications à la conception ensembliste et au tolérancement, PhD Thesis, École Nationale Supérieure d'Arts et Métiers, 2011.

[QUI 12] QUIRANTE T., Modeling method and numerical optimization for decision support in robust embodiment design of products and processes, PhD Thesis, University of Bordeaux, 2012.

[QUI 13] QUIRANTE T., SEBASTIAN P., LEDOUX Y., "A trade-off function to tackle robust design problems in engineering", *Journal of Engineering Design*, vol. 24, no. 1, pp. 64–81, 2013.

[RAJ 93] RAJASHEKHAR M.R. ELLINGWOOD B.R., "A new look at the response surface approach for reliability analysis", *Structural Safety*, vol. 12, no. 3, pp. 205–220, 1993.

[RAN 08] RANJAN P., BINGHAM D., MICHAILIDIS G., "Sequential experiment design for contour estimation from complex computer codes", *Technometrics*, vol. 50, no. 4, pp. 527–541, 2008.

[RAS 94] RASMUSSEN J., PEJTERSEN A.M., GOOSTEIN L.P., *Cognitive Systems Engineering*, John Wiley & Sons, New York, 1994.

[RES 78] RESNICK S.I., *Extreme Values, Regular Variation and Point Processes*, Springer-Verlag, Berlin, 1978.

[RIC 08] RICAUD J.L., "Allocution d'ouverture du président du comité d'organisation", *15th Congress* $\lambda\mu$, Lille, IMdR, 2008.

[ROC 02] ROCCO C.M., MORENO J.A., "Fast Monte Carlo reliability evaluation using support vector machine", *Reliability Engineering & System Safety*, vol. 76, no. 3, pp. 237–243, 2002.

[ROM 04] ROMERO V.J., SWILER L.P., GIUNTA A.A., "Construction of response surfaces based on progressive lattice-sampling experimental designs with application to uncertainty propagation", *Structural Safety*, vol. 26, no. 2, pp. 201–219, 2004.

[RUB 81] RUBINSTEIN R.Y., *Simulation and the Monte Carlo Methods*, John Wiley & Sons, New York, 1981.

[RUB 08] RUBINSTEIN R.Y., KROESE D.P., *Simulation and the Monte Carlo Method*, Wiley Series in Probability and Statistics, Wiley, 2008.

[RYC 05] RYCKELYNCK D., "A priori hypereduction method: an adaptive approach", *Journal of Computational Physics*, vol. 202, pp. 346–366, 2005.

[RZH 49] RZHANITZYN R.A., "Design of structures with considerations of plastic properties of materials", *Stroivoenmorizdat*, Moscow, 1949.

[SCO 98] SCOTT M.J., ANTONSSON E.K., "Aggregation functions for engineering design trade-offs", *Fuzzy Sets and Systems*, vol. 99, pp. 253–264, 1998.

[SAA 01] SAATY T.L., *Fundamentals of Decision Making and Priority Theory*, AHP series, RWS publications, vol. 6, pp. 478, 2001.

[SAA 06] SAATY T.L., "Rank from comparisons and from ratings in the analytic hierarchy/network processes", *European Journal of Operational Research*, vol. 168, pp. 557–570, 2006.

[SAL 04] SALTELLI A., TARANTOLA S., CAMPOLONGO F., *et al.*, *Sensitivity Analysis in Practice – A Guide to Assessing Scientific Models*, John Wiley & Sons, New York, 2004.

[SAL 08] SALTELLI A., RATTO M., ANDRES T., *et al.*, *Global Sensitivity Analysis: The Primer*, Wiley, 2008.

[SAL 10] SALTELLI A., ANNONI P., AZZINI I., *et al.*, "Variance based sensitivity analysis of model output. design and estimator for the total sensitivity index", *Computer Physics Communications*, vol. 181, pp. 259–270, 2010.

[SAR 08] SARIDAKIS K.M., DENTSORAS A.J., "Soft computing in engineering design – a review", *Advanced Engineering Informatics*, vol. 22, pp. 202–221, 2008.

[SCH 03] SCHLATHER M., TAWN J.A., "A dependence measure for multivariate and spatial extreme values distributions: properties and inference", *Biometrika*, vol. 90, pp. 139–156, 2003.

[SCA 04] SCARAVETTI D., Formalisation préalable d'un problème de conception, pour l'aide à la décision en conception préliminaire, PhD Thesis, ENSAM Bordeaux, 3 December 2004.

[SCH 08] SCHILDERS W.H.A., VAN DER VORST H.A., ROMMES J., *Model Order Reduction: Theory, Research Aspects and Applications*, Mathematics in Industry, vol. 13, Springer, 2008.

[SÉB 12] SÉBASTIAN P., LEDOUX Y., COLLIGNAN A., *et al.*, "Linking objective and subjective modeling in engineering design through arc-elastic dominance", *Expert Systems with Applications*, vol. 39 no. 9, pp. 7743–7756, July 2012.

[SHA 76] SHAFER G., *A Mathematical Theory of Evidence*, Princeton University Press, 1976.

[SHA 78] SHAFER G., "Non-additive probabilities in the work of Bernoulli and Lambert", *Archive for History of Exact Sciences*, vol. 19, pp. 309–370, 1978.

[SHA 86] SHAFER G., "The combination of evidence", *International Journal of Intelligent Systems*, vol. 1, no. 3, pp. 155–179, 1986.

[SHA 95] SHARPE J.E.E., "Computer tools for integrated conceptual design", *Design Studies*, vol. 16, no. 4, pp. 471–488, 1995.

[SIM 69] SIMON H.A., *The Sciences of the Artificial.*, 1st ed., MIT Press, Cambridge, MA., 1969.

[SIM 82] SIMON H.A., *Models of Bounded Rationality*, MIT Press, Cambridge, MA, vol. 2, 1982.

[SMO 04] SMOLA A.J., SCHLKOPF B., "A tutorial on support vector regression", *Statistics and Computing*, vol. 14, no. 3, pp. 199–222, 2004.

[SOI 00] SOIZE C., "A nonparametric model of random uncertainties for reduced matrix models in structural dynamics", *Probabilistic Engineering Mechanics*, vol. 15, pp. 277–294, 2000.

[SOI 04] SOIZE C., GHANEM R., "Physical systems with random uncertainties: chaos representations with arbitrary probability measure", *SIAM Journal of Scientific Computing*, vol. 26, no. 2, pp. 395–410, 2004.

[SOI 05] SOIZE C., "A comprehensive overview of a non-parametric probabilistic approach of model uncertainties for predictive models in structural dynamics", *Journal of Sound and Vibration*, vol. 288, no. 3, pp. 623–652, 2005.

[SRI 99] SRINIVASAN V., "On interpreting key characteristics", *ASME Design Engineering Technical Conferences*, Las Vegas, NV., 12–15, September 1999.

[STE 09] STEFANOU G., "The stochastic finite element method: past, present and future", *Comput. Methods Appl. Mech. Engrg.*, vol. 198, pp. 1031–1051, 2009.

[SUD 07a] SUDRET B., "Global sensitivity analysis using polynomial chaos expansion", *Reliability Engineering and System Safety*, vol. 93, pp. 964–979, 2007.

[SUD 07b] SUDRET B., Uncertainty propagation and sensitivity analysis in mechanical models – Contributions to structural reliability and stochastic spectral methods, Habilitation à diriger des recherches, Blaise Pascal University, Clermont-Ferrand, France, 2007.

[TAG 89] TAGUCHI G., *Introduction to Quality Engineering*, American Supplier Institute, 1989.

[TAL 07] TALEB N.N., *The Black Swan: The Impact of the Highly Improbable*. Random House, 2007.

[TER 99] TERWIESCH C., LOCH C.H., "Measuring the effectiveness of overlapping development activities", *Journal of Management Science*, vol. 45, no. 4, pp. 455–465, 1999.

[THA 04] THACKER B.H., DOEBLING S.W., HEMEZ F.M., *et al.*, Concepts of model verification and validation, Technical Report LA-14167-MS, Los Alamos National Laboratory, October 2004.

[THÉ 03] THÉSAME., Available at: http//www.thesame-innovation.com/publi/publi.php mécatronique: la mécatronique à l'épreuve du marché, 2003.

[THU 03] THUNNISSEN D.P., "Uncertainty classification for the design and development of complex systems", *3rd Annual Predictive Methods Conference*, Newport Beach, CA, June 2003.

[TOL 98] TOLLENAERE M., *Conception de produits mécaniques*, Hermes, Paris, 1998.

[VAP 95] VAPNIK V., *The nature of Statistical Learning Theory*, Springer-Verlag New York, 1995.

[WIE 38] WIENER N., "The homogeneous chaos", *American Journal of Mathematics*, vol. 60, pp. 897–936, 1938.

[XIU 02] XIU D., KARNIADAKIS G.E., "Modeling uncertainty in steady state diffusion problems via generalized polynomial chaos", *Computer Methods in Applied Mechanics and Engineering*, vol. 191, pp. 4927–4948, 2002.

[XIU 03] XIU D., KARNIADAKIS G.E., "Modelling uncertainty in flow simulations via generalized polynomial chaos", *Journal of Computational Physics*, vol. 187, pp. 137–167, 2003.

[YAS 99] YASSINE A., FALKENBUR D., CHELST K., "Engineering design management: an information structure approach", *International Journal of Production Research*, vol. 37, no. 13, pp. 2957–2975, 1999.

[YUA 09] YUAN X.K., LU Z.Z., LU Y.B., "Support vector machine response surface method based on fast Markov chain simulation", *Proceedings of 2009 IEEE International Conference on Intelligent Computing and Intelligent Systems (ICIS 2009)*, Shanghai, vol. 4, pp. 279–282, 2009.

[ZAD 65] ZADEH L.A., "Fuzzy sets", *Information and Control*, vol. 8, pp. 338–353, 1965.

[ZAD 78] ZADEH L.A., "Fuzzy sets as a basis for a theory of possibility", *Fuzzy Sets and Systems*, vol. 1, pp. 3–28, 1978.

[ZEY 13] ZEYDINA O., BEAUZAMY B., *Probabilistic Information Transfer*, Société de Calcul Mathématique S.A., 2013.

Index

G, H, I

K, L

M

O

P